Praise for *Thanks, Obama*

New York Times **Bestseller •** ***Washington Post*** **Bestseller**
IndieBound Bestseller • An ***Esquire*** **Best Book of the Year**

"David Litt has done the impossible: written a smart, insightful, and funny White House memoir you don't have to be a political junkie to love."
—Judd Apatow

"Litt is a funny and skillful storyteller. . . . While the first half of the book is enjoyable, the second half is masterly, rising to a crescendo that is as rousing as, well, a particularly inspiring campaign speech."
—*New York Times Book Review*

"Genuinely charming, and perceptive. . . . Graceful, instructive, wry speechwriter memoirs like Litt's are the exception rather than the norm. I think *Thanks, Obama* will join the ranks of lasting works about the culture and texture of political life, and of coming-of-age accounts by staffers who grow up personally and politically at the same time."
—*The Atlantic*

"Serve[s] as a more devastating indictment of the current administration than a campaign-style book ever could. . . . Limber, funny and illuminating."
—*New Republic*

"A thoughtful and funny account of life as a minnow surrounded by Washington's self-important whales. . . . Ranks with other classics from former White House speechwriters, such as Peggy Noonan's *What I Saw at the Revolution.*"
—*USA Today*

"Irresistibly charming. . . . Litt minted his star converting world affairs into jokes. The translation of satire back to sincerity is trickier to pull off, and lands with its own undeniable grace."
—Slate

"By turns moving and hilarious, David Litt's rollicking account of his journey from campaign field grunt to presidential speechwriter is an irresistible read."
—David Axelrod, former senior advisor to Barack Obama and author of *Believer: My Forty Years in Politics*

"Offers both humor and optimism, two things many of us sorely need these days." —Bustle, Best of the Month

"Highly entertaining. . . . Much more than a scrapbook of Beltway gossip and Obama idolatry." —*Pacific Standard*

"His time [in the White House] was as 'hopey changey' as advertised—with a little bit of absurdity and humor added into the mix."
—*Elle,* Best of the Month

"Highlights the power of presidential rhetoric, via the stark differences between how it is used now versus in the Obama administration. . . . What Litt understands and what *Thanks, Obama* makes clear may very soon be forgotten: The finest presidential speeches can heal the nation."
—Paste

"Terrific—part firsthand story about being inspired by a cultural icon, part how-to manual for getting involved in politics and making change. *Thanks, Obama* is a hysterical, pithy, and heartfelt trip down memory lane." —Keegan-Michael Key

"Funny and unexpectedly moving. . . . A powerful reminder that true fulfillment can come from wielding even the smallest bit of influence on behalf of those who have none." —*Washington Monthly*

"David Litt is brilliant. . . . Always intelligent, razor sharp, and hilarious." —Billy Eichner

"Don't be fooled by the self-deprecating narrator; this portrait of a young speechwriter is filled with massive wit, hard-earned wisdom, and a loving touch. David's labors remind us of a not-so-distant past when words mattered." —Matt Walsh, HBO's *Veep*

"Down-to-earth and amusing. . . . Shows us the journey of a young man coming into his own, politically, personally, and professionally."
—*Book Riot*, Best of the Month

"Reminds us that a great president galvanizes not only his staff but his country." —Anne Fadiman, author of *Ex Libris*

THANKS,
OBAMA

(*From left to right:*) Not me; me; also not me.

THANKS, OBAMA

MY HOPEY, CHANGEY WHITE HOUSE YEARS

DAVID LITT

ecco

An Imprint of HarperCollins*Publishers*

For Jacqui,
who sees right through me
and likes me anyway

Photograph on page iv courtesy of Lawrence Jackson.

For information, address HarperCollins Publishers, 195 Broadway, New York, NY 10007.

HarperCollins books may be purchased for educational, business, or sales promotional use. For information, please e-mail the Special Markets Department at SPsales @harpercollins.com.

A hardcover edition of this book was published in 2017 by Ecco, an imprint of HarperCollins Publishers.

FIRST ECCO PAPERBACK EDITION PUBLISHED 2018.

Designed by Renata De Oliveira

Library of Congress Cataloging-in-Publication Data has been applied for.

ISBN 978-0-06-256844-1

18 19 20 21 22 LSC 10 9 8 7 6 5 4 3 2 1

CONTENTS

A NOTE REGARDING FACTS

(on the theory that they still exist)

I didn't keep a diary while working at the White House—I didn't want anything Congress could subpoena. This book is based on memory, vigorous Googling, and stories told and retold to family and friends. To the extent possible, it has been professionally fact-checked. Every quotation includes as much verbatim language as I can recall and preserves, at the very least, the intention of what was said. (I tried to be particularly careful when those quotes are President Obama's.) For public figures, I used real names; for most others, I used pseudonyms.

Introduction

ARUGULA ON AIR FORCE ONE

"That motherfucker be sliding!"

The guy leaning out his car window doesn't know he's shouting at the motorcade. Nor, in all likelihood, would he care. It is January 20, 2016, and an inch of snow has fallen on the District of Columbia. That's more than enough to throw the nation's capital into chaos. We're halfway between *Frozen* and *Mad Max*.

Presidents aren't supposed to get stuck in traffic. That's one of the job's best perks. Tonight, however, is an exception. The same snowstorm that emerged from nowhere to snarl Washington's roads grounded the president's helicopter. There wasn't even time to clear a path for his car. The best the military office could offer was an upgrade. Ordinarily POTUS travels in "The Beast," a tank wearing a limousine costume, but with ice on the ground, thick armor plating has been traded in for traction. Barack Obama is still commander in chief. Markets move on his decisions. Nations can be decimated at his command. But tonight, Barack Obama is also just another middle-aged dad in an SUV, struggling to make it home on time from work.

At least he has four-wheel drive. Junior staffers like me are in ordinary fifteen-passenger vans. We're fishtailing like crazy.

I had not expected to leave Andrews Air Force Base and run straight into a metaphor, but that's what happened. Washington is hopelessly gridlocked. We're moving forward more slowly than anyone would like. It seems only fitting my last-ever POTUS trip would end this way, confident we're heading in the right direction but concerned the wheels are coming off the bus. As we carom toward a bank of parked cars, I can even hear our self-appointed pundit deliver a fresh critique.

"That motherfucker be *sliding* right now!"

Against all odds, we regain control. Our incremental progress continues.

When I boarded the plane that morning, I was thinking less about symbolism and more about snacks. There was a time when entering Air Force One was like stepping through a closet into Narnia. By my final flight, however, I had developed a routine. Climb the stairs, walk past the conference room, pluck a handful of grapes from the fruit bowl. Hang my jacket in the closet, grab an Ethernet cable, swipe a box of presidential M&M's. Order an iced coffee, deploy the retractable footrest, put on the enameled metal pin reminding Secret Service agents not to shoot me. Then try to finish editing my speech before lunch.

Whenever I saw POTUS eating on the plane it was something healthy, usually just a chicken breast and veggies. The rest of us ate food I can only assume was prepared by cannibals fattening us up. The meals were packed with calories, the menus with adjectives. On that morning's flight out of Andrews, a short one to Detroit, we'd been served creamy Brie cheese with crispy pancetta on toasted rustic garlic bread. The fresh arugula had been topped with fresh cracked pepper and shaved Parmesan cheese.

I once brought this up with Ted, a crew member. Why were even the "lighter options" covered in bacon bits or doused with melty cheddar?

"An army marches on its stomach," he replied.

That might be true for actual armies, ones with soldiers who march long distances and burn calories killing people. As a speechwriter, I didn't march. Enemy bullets were not a concern. Food coma was. Aboard the presidential aircraft, I ate stuffed pork chops and crab pretzels and giant cups of buffalo blue cheese dip that were, remarkably, categorized as snacks. After a last-minute edit, I'd reward myself with fun-size Twix or Snickers from the candy tray by the window. Then there were the actual desserts. Who knows how many pecan pies and strawberry parfaits, apple tarts and brownies à la mode I polished off in service to my country?

IF YOU'D ASKED ME TEN YEARS EARLIER WHAT I MIGHT BE DOING AT age twenty-nine, clogging my arteries on Air Force One would not have made the list. True, I went to Yale, the kind of fancy-pants university where a sizable number of students have been running for office since birth. But not me. I imagined spending my twenties squeezing every drop of adventure from life. I would trek through far-flung landscapes and learn new languages and develop six-pack abs. I would disrupt institutions. I would subvert them or transcend them. But join them? Never. That would be pathetic.

Fast-forward a decade. I have taken zero journeys of self-discovery, but own a robust menagerie of ties. I carry a thin stack of business cards in my wallet and a thicker stack of backup cards in my bag. Each time I fly for work, an Air Force officer hands out warm towels and addresses me, without irony, as "sir."

When I'm not careful, I even start to think I deserve it.

But events have a way of cutting staffers down to size. Two months before my Detroit trip, I went to see President Obama record his weekly address. I usually hid discreetly in the corner for these tapings, but this time, for reasons that now escape me, I sat front and center. When POTUS glanced toward the teleprompter, we accidently locked eyes.

Few activities offer less upside than a staring contest with the president. But now, having started one, I didn't know how to stop. I considered averting my gaze, like a shy maiden in a Jane Austen novel, but that would only make things more awkward. I kept looking at President Obama. President Obama kept looking at me. Finally, after what seemed like hours, he spoke.

"What are *you* doing here?" He wasn't annoyed, exactly. He just seemed to find my presence unexpected, the way you might be surprised to discover your dog in the living room instead of in its crate.

A different young staffer would have handled the situation gracefully. Perhaps they might have tried a high-minded approach: "I'm here to serve my country." Or they might have kept things simple: "I'm hoping to catch typos."

Here is what I did instead. First, in a misguided effort to appear casual, I gave the leader of the free world a smile reminiscent of a serial killer who knows the jig is up. Then I said the following:

"Oh, I'm just watching."

POTUS took a shallow breath through his nose. He raised his eyebrows, looked at our cameraman, and sighed.

"It always makes me nervous when Litt's around."

I'm 90 percent sure President Obama was half joking. Still, two months later, on my final POTUS trip, my stomach full of arugula and Brie, I was careful to avoid his eyes. Backstage in Detroit, POTUS went through his usual prespeech routine, shaking hands with the prompter operators and joking with personal aides. Then he stepped onstage to remind a roomful of autoworkers about the time he saved their industry seven years before.

I had written plenty of auto speeches for President Obama. There was nothing especially new in this one. But as POTUS reached his closing paragraph, my eyes filled with tears. I had tried to prepare myself for each milestone: my last set of remarks for the president, my last ride in the motorcade, my last flight on Air Force One. Still, the nostalgia left me reeling. I fled the staff viewing area and found a men's room. With my left hand, I steadied myself against the sink. With my right, I held all but the first page of my speech.

You're supposed to be an adult, I reminded myself. *And adults don't cry in front of their boss's boss.*

I pulled it together, took a deep breath, and returned to the hold room to wait. Presidential trips are like that. One moment your fortunes are tied, inextricably, to the most important person on earth. The next moment you're killing time in someone's abandoned third-grade classroom or empty office suite. Five minutes passed. Ten minutes. Then a voice rang out from the hall.

"Litt!"

It was POTUS. With his left hand he clutched the first page of my speech, now inscribed with his unmistakable signature. He held his right hand palm up, for a shake.

"You didn't tell me you were leaving," he said.

"Well, actually, I'm trying to sneak out." By my low standards, this was remarkably good banter. The president bantered back.

"You didn't do a very good job. I caught you."

He started to ask a question, but one of his aides gestured toward cameras set up for a post-speech interview. "Never mind," he said. "We'll talk more on the plane."

We didn't, of course. On the flight home, the president was busy being the president, and I was busy eating Cuban picadillo with a fresh side salad and keeping my feelings at bay. It wasn't until we were about a half hour from Andrews Air Force Base that I heard the phrase "bad weather call." Not long after that we landed in the snowstorm. Not long after that, we loaded the motorcade only to find every inch of asphalt choked with cars.

And now we're going nowhere. The light turns red. The motorcade once again grinds to a halt, this time beside a Chick-fil-A. Another metaphor. I'm frustrated and nervous, wondering if anyone has a plan.

On cue, Sarah Palin's voice pops into my head. She's always doing this, showing up when my spirits are lowest. It's like I have a fairy godmother who hates me.

"So," she asks, "how's that whole hopey, changey thing workin' out for ya?"

It's a line she started using in 2010, when President Obama's approval ratings were plummeting and the Tea Party was on the rise. And here's the thing: if you ignore her mocking tone and that annoying dropped *G*, it's a good question. I spent the lion's share of my twenties in Obamaworld. Career-wise, it went well. But more broadly? Like so many people who fell in love with a candidate and then a president, the last eight years have been an emotional roller coaster. Groundbreaking elections marred by midterm shellackings. The exhilaration of passing a health care law followed by the exhaustion of defending it. Our first black president made our union more perfect simply by entering the White House, but a year from now he'll vacate it for Donald Trump, America's imperfections personified.

The motorcade keeps skidding and sliding. For twenty miles

we veer left and right, one close call after another, until we finally reach the South Lawn. Here, too, I have a routine: get out of the van, walk through the West Wing, head to my office across the street. It's a trip I've made countless times before. It's also one I will never make again. And as I walk past the Rose Garden, the flagstones of the colonnade pressing against the soles of my leather shoes, Sarah Palin's question lingers in the January air.

How has it all worked out?

PART ONE

OBAMABOT

1

THE RAPTURE

On January 3, 2008, I pledged my heart and soul to Barack Obama. There was no formal, lovesick declaration. No one tattooed a Hope poster across my chest. Still, my transformation was immediate and all-consuming. One moment I was a typical college senior, barely interested in politics. The next moment I would have done anything, literally anything, for a freshman senator from Illinois.

I was not a likely candidate for conversion. The summer before I began working for Obama I interned at the comedy newspaper *The Onion,* where my boss wore roller-skate sneakers and sold feminine hygiene products from a kiosk at his desk. It was a dream job. I fetched coffee and did busy work. In exchange, I got to sit in on a writers' meeting and watch a senior editor come dangerously close to a psychotic break. "We're a comedy paper, not a stupid paper!" he shouted, before storming out of the room. I had never been part of anything so meaningful.

There was just one problem: I didn't fit in. As an intern, my biggest responsibilities were proofreading articles and writing jokes about the weather, but the second task kept getting in the way of the first. Each morning, I'd arrive at work and think, *Cloudy with a chance of meatballs!* I knew it wasn't funny, but the phrase lodged itself in my head like a mantra, or a tumor. Typos went uncorrected. Run-on sentences ran on.

Cloudy with a chance of meatballs. Cloudy with a chance of meatballs. Cloudy with a chance of meatballs.

This wasn't just another job. I worshipped *The Onion.* I grew up in Manhattan, and I'll never forget the headlines from the issue released a few weeks after 9/11, when I still thought al Qaeda would kill me before I finished tenth grade.

HIJACKERS SURPRISED TO FIND SELVES IN HELL

NOT KNOWING WHAT TO DO,
WOMAN BAKES AMERICAN-FLAG CAKE

In that awful moment, a small, satirical newspaper was everything I loved about my country. Defiant. Proud. Optimistic in spite of everything. *The Onion* gave me hope I might not die a virgin. What could be more uplifting than that?

But if satire represented the best of America, politics was the worst. My family is a classic American-dream story. My great-grandparents fled Russia to avoid being murdered for their religion. Just two generations later, my parents fled New York City weekends for their country house. I never felt guilty about this. I was raised to believe America rewards hard work. But I was also raised to understand that luck plays a role in even the bootstrappiest success story. The cost of living the dream, I was taught, is the responsibility to expand it for others. It's a more than fair price.

Yet the people running the country didn't see it that way. With George W. Bush in the White House, millionaires and billionaires were showered with tax cuts. Meanwhile, schools went underfunded. Roads and bridges deteriorated. Household incomes languished. Deficits ballooned.

And America went to war. President Bush invaded Iraq to destroy weapons of mass destruction, a campaign which hit a snag when it turned out those weapons didn't exist. But by then it was too late. We had broken a country and owned the resulting mess. Colin Powell called this "the Pottery Barn rule," which, admittedly, was cute. Still, it's hard to imagine a visit to Pottery Barn that costs trillions of dollars and thousands of American lives.

Our leaders, in other words, had made bad choices. They would therefore be replaced with better ones. That's how AP Government told me the system worked. In the real world, however, the invasion of Iraq became an excuse for a dark and antidemocratic turn. Those who questioned the war, the torture of prisoners—or even just the tax cuts—found themselves accused of something barely short of treason. No longer was a distinction made between supporting the president's policies and America's troops. As an electoral strategy, this was dangerous and cynical. Also, it worked.

So no, I didn't grow up with a high opinion of politicians. But I did grow up in the kind of environment where people constantly told me I could change the world. In 2004, eager to prove them right, I volunteered for John Kerry's presidential campaign.

In theory, we stood on the right side of history. For equality! For opportunity! For the little guy! In practice, however, being branded un-American left Democrats meek and skittish, like the Munchkins before Dorothy arrives. I had no doubt Kerry would make a better president than Bush, yet he never seemed confident when stating his case. It was as though he spent an entire campaign arguing that the most talented Beatle was Ringo. When he lost, I was devastated. More than that, however, I was embarrassed. I had allowed myself to believe my meager actions could alter a country's course. How foolish that seemed now. How naive.

I was done with politics. And I was through believing in clichés. "Changing the world" was for hypocrites, the kind of people who were outraged by a nonorganic tomato but never asked questions about their weed. "Taking our country back" was for budding white-collar criminals who wore suits and ties to class.

And me? Once I realized I couldn't change the world, I doubled down on making fun of it. My greatest passion in college was my improv comedy group. My second-greatest passion was a humor magazine. When I arrived at *The Onion* and discovered that my happiest coworkers were goofy, awkward nihilists, I wasn't disenchanted. I was thrilled. I longed to be charmingly bitter. I dreamed of one day melting down in meetings before storming out of rooms. I was determined to write the best gosh-darned jokes about the weather the paper had ever seen!

Cloudy with a chance of meatballs. Cloudy with a chance of meatballs. Cloudy with a chance of meatballs.

It can be hard, at times, to distinguish between the absence of talent and the presence of destiny. When I began my dream job, I imagined buying a wholesale tub of maxipads and following in my boss's footsteps, or, if his skates were deployed, his tracks. But when August rolled around, my fellow intern Mariana had landed about six jokes in the paper. I had landed about none.

You know, I thought, *maybe this job isn't so meaningful after all.*

For the first time in my life, I was seeking a higher purpose, but after my experience with the Kerry campaign, politics never crossed my mind. Instead, I applied to join the CIA. With my major in history and leadership experience directing my comedy troupe, I figured I was the perfect person to bring Osama bin Laden to justice.

I don't remember where I was when the CIA called, although since it was my senior year of college, I was probably either recovering from a hangover or acquiring one. I also don't remember my interviewer's name. I do, however, recall that it was something all-American, like Chip or Jimmy. I also remember that he sounded surprisingly sunny, as though he were selling time-shares or cutlery door-to-door.

"Alrighty now," said Buddy, or maybe even Tex. "Just to kick things off, have you used any prohibited substances in the past year?"

If I had lied to the CIA, perhaps I might have passed a test. Instead of writing a book about the White House, I'd be poisoning a drug kingpin with a dart gun concealed inside a slightly larger dart gun, or making love to a breathy supermodel in the interest of national security. I'll never know. I confessed to smoking pot two months before.

The sunniness vanished from my interviewer's voice. "Normally we like people who break the rules," Skipper told me, "but we can't consider anyone who's used illegal substances in the past twelve months." Just like that, my career as a terrorist hunter was over.

I thought my yearning for higher purpose would vanish with my CIA dreams, the way a Styrofoam container follows last night's Chinese food into the trash. To my surprise, it stuck around. In the weeks that followed, I pictured myself in all sorts of identities: hipster, world traveler, banker, white guy who plays blues guitar. But these personas were like jeans a half size too small. Trying them on gave me an uncomfortable gut feeling and put my flaws on full display. My search for replacement selves began in November. By New Year's Eve I was mired in the kind of existential funk that leads people to find Jesus, or the Paleo diet, or Ayn Rand.

Instead, on January 3, I found a candidate.

I was on an airplane when I discovered him, preparing for our initial descent into JFK. This was during the early days of live in-flight television, and I was halfway between the Home Shopping Network and one of the lesser ESPNs when I stumbled across coverage of a campaign rally in Iowa. Apparently, a caucus had just finished. Speeches were about to begin. With nothing better to occupy my time, I confirmed that my seat belt was fully fastened. I made sure my tray table was locked. Then, with the arena shrunk to fit my tiny seatback screen, I watched a two-inch-tall guy declare victory.

It's not like I hadn't heard about Barack Obama. I had heard his keynote speech at the 2004 Democratic Convention. His presidential campaign had energized my more earnest friends. But I was far too mature to take them seriously. They supported someone with the middle name *Hussein* to be president of the United States. While they were at it, why not cast a ballot for the Tooth Fairy? Why not nominate Whoopi Goldberg for pope?

And then I saw him speak.

Years later, after writing dozens upon dozens of presidential speeches, it would become impossible to listen to rhetoric without editing it in my head. On that historic Iowa evening, Obama began with a proclamation: "They said this day would never come." Rereading those words today, I have questions. *Who were "they," exactly? Did they really say "never"?* Because if *they* thought an antiwar candidate with a robust fund-raising operation could never win a divided three-way Democratic caucus, particularly with John Edwards eating into Hillary Clinton's natural base of support among working-class whites, then *they* didn't know what *they* were talking about.

All this analysis would come later, though, along with stress-induced insomnia and an account at the Navy Mess. At the time, I was spellbound. The senator continued:

"At this defining moment in history, you have done what the cynics said you couldn't do." He spoke like presidents in movies. He looked younger than my dad. I didn't have time for a second thought, or even a first one. I simply believed.

Barack Obama spoke for the next twelve minutes, and except for a brief moment when the landing gear popped out and I thought we were going to die, I was riveted. He told us we were one people. I nodded knowingly at the gentleman in the middle seat. He told us he would expand health care by bringing Democrats and Republicans together. I was certain it would happen as he described. He looked out at a sea of organizers and volunteers.

"You did this," he told them, "because you believed so deeply in the most American of ideas—that in the face of impossible odds, people who love this country can change it."

Like most twenty-one-year-olds, I was no stranger to the sudden, all-consuming crush. "There's this girl," I would gush to friends who tolerated that sort of thing. "She's from California, and I once spent a week in Washington State! Can you believe how much we have in common?" Watching Obama speak, my attraction was electoral rather than physical. But in politics, as in other things, the heart wants what it wants.

I do *love this country!* I thought. *I* can *change it! It's like he's known me my whole life!*

As we neared the runway, I tried to make sense of what had just happened. I was born in the tail end of the Reagan years, when government was not the solution but the problem. I cast my first vote during the Bush years, when "You are either with us or with the terrorists" was applied to foreign and domestic opponents alike. Now, a few thousand feet over New York City, a candidate for president had told me we were not a collection of red states and blue states, but the United States. Together, we could build something far greater than we could on our own.

By the time we emerged from the Jetway, I was one of those people who would not shut up about Barack Obama. I wasn't alone. Across campus, across America, an army of idealists had arisen, a zombie horde craving hope and change.

Our critics would later mock the depths of our devotion. Obamabots, they'd call us. And really, weren't they right? Becoming obsessed with Barack Obama wasn't a choice I made. Rather, it was like starring in one of those sleeper-agent-killer-robot movies that comes out every few years. A switch is flipped, long-dormant code is activated, and suddenly the mild-mannered main character can disembowel adversaries with a spoon. I've never disemboweled anybody, not even people who actually use the phrase, "Find me on LinkedIn." Still, I identify with that killer robot. I had been preprogrammed with the ability to ask friends for donations or to call people at random to tell them how to vote. Now, my switch had been flipped.

When I got back to campus, I joined our chapter of Obama for America. Organizers handed out call sheets, pieces of paper covered in strangers' numbers and names, and each night I dialed until my fingers were sore. These days I'm more likely to receive these calls than make them. I hang up so quickly, you'd think someone was trying to poison me over the phone. But in 2008, that unicorn of political seasons, Democrats were happy to take unsolicited advice from a stranger who had been legally drinking for all of four months.

I came to think of men and women I cold-called as "my voters." If they didn't pick up, I'd leave a helpful, minute-long voice mail. If they did pick up, I'd deliver the exact same message, only with room for questions at the end. I tailored my pitch in small ways. Tiffany might hear about Obama's ability to bring people together. Tucker might hear about his midwestern roots. Treshawn might hear the word *historic* mentioned three or four times in a single sentence.

Mostly, though, I talked about Iraq. On the eve of the war, Hillary Clinton voted to give George W. Bush authority to invade. Was she motivated by principle? By her desire to seem tough? No one could tell. Contrast that with Obama. In 2002, when opposing Bush was political suicide, he called Iraq a "dumb war." Being president took two things, I told my voters: judgment and courage. With just one speech, my candidate had demonstrated both.

"What about experience? Hasn't he only been in the Senate for, like, two years?"

"I don't think that matters," I assured them. I would have set myself on fire before allowing a sophomore to direct my improv comedy group. When it came to running the country, however, I was pretty sure a freshman senator could figure it out.

I graded my voters on a one-to-five scale. Fives supported Hillary. Threes were undecided. Ones supported us. In a week or two of phone calls I covered almost every imaginable issue: electability, education, infrastructure, GMOs. The secret to these conversations, I learned, was to substitute personal detail for genuine expertise.

"Of course I support America's farmers. I eat salad all the time!"

"As someone with four grandparents, I can't afford *not* to worry about social security!"

That sort of thing.

There was one issue my fellow Democrats and I rarely discussed but was always on our minds. Race. I was calling on behalf of an African American candidate who had won in Iowa, where the electorate was more than 90 percent white. This was impossible. Yet it had been done. And that was at the heart of my candidate's appeal. Obama wasn't just fighting for change. He *was* change. He was the messenger and message all at once. It's one thing to follow a prophet who speaks glowingly of a promised land. It's another thing entirely to join him once he parts the sea.

Given the circumstances, it seemed selfish *not* to spread the good news. Overnight, my friends found themselves living with an evangelist in their midst, an Obama's Witness who could take or leave your soul but was desperate for your vote. When it came to tactics, I took my inspiration from the heroes who came before me. Mahatma Gandhi went on a hunger strike. Rosa Parks refused to yield her seat. I pasted my entire address book into a "bcc" field and wrote my take on that week's news.

"Obama's big wins in the last few days are largely thanks to the number of supporters he has getting out the vote for him, and every landslide victory gets him even more momentum!"

No one replied to my bulletin, but that didn't bother me. I loved my country. I was changing it. Besides, things had probably been just as difficult for Martin Luther King.

FOUR WEEKS AFTER MAKING MY FIRST CALL FOR BARACK OBAMA, I got the chance to vote for him: Connecticut held its primary on February 5. We were a small state, but our position next to Hillary Clinton's adopted home of New York gave us outsize importance. About a week before the election, the campaign announced that Obama himself would come to deliver a speech.

Like most Jews, I haven't spent much time waiting for the rapture. But after the nights I lay awake, counting down the minutes until that rally, I think I get the appeal. On February 4, when the anointed hour arrived, I gathered a crew of fellow supporters and borrowed my roommate's car without her permission. Then we made the pilgrimage to Hartford, Connecticut, our temporarily sacred ground.

A few years later, when I traveled to rallies in the motorcade, I would sometimes wonder why anyone in the audience would want to go. Hours before the speaker says a word, you wait for the doors to open. You wait to go through metal detectors. You wait for the program to begin. You wait for the speaker to speak. After at least two hours of waiting and at most one hour of speaking, you wait for the speaker to exit. Then you wait for everyone ahead of you to exit. Then, after all that, you wait for your bus or train or car. Presidential speeches are decathlons of standing around.

Why not just watch online? I think, forgetting about that day in Hartford when I went myself, and waited for hours, and would not have traded a single second away. On the floor of a basketball arena, surrounded by sixteen thousand fellow pilgrims, we hoisted homemade signs. We did the wave. A few of us tried to spark a cheer.

"Yes we can! Yes we can! Yes we . . . not yet? Okay, never mind."

Finally, long after we lost track of time, the program began. Ted Kennedy and Barack Obama were wrapping up a week of joint appearances, a political odd couple on the road trip of their lives. It was like watching Julie Andrews and Lady Gaga team up for a Christmas album. The veteran had lost a step, but got by on decades of accumulated wisdom. The newcomer was raw at times, but possessed a talent that could not be denied. Together, they covered all the Democratic standards. Ending the Iraq War. Affordable health care. Fighting for the middle class.

Yet here's the remarkable thing: I don't remember a word. On that plane into JFK, I was captivated by the candidate. In Hartford, I was captivated by the crowd. I had seen diversity before, of course, on the front pages of college admissions brochures. But looking up at thousands of screaming Democrats, I realized I was part of a truly diverse group of people for the first time in my life.

It's always risky to reduce American society to "One Fish, Two Fish, Red Fish, Blue Fish," but there was no tokenism in the arena that afternoon. Young people, old people. Gay people, straight people. Black people, white people. Men and women. Rich and poor. So often in America, these differences were dividing lines, but here in Hartford lay the promise of something better, personified not by our candidate, but by us. We were proud of who we were and where we came from. Most of all, though, we were proud to be part of something bigger than ourselves.

"U-S-A! U-S-A!"

I heard the cheer billowing behind me. To my surprise, I joined in. This kind of raw patriotism had been co-opted during the Bush years, when we were told the only way to love your country was to support invading another one. The Right had claimed words like *freedom* and *liberty,* words the Left was all too willing to abandon. Now here we were, pumping our fists without irony.

Nor were we cheering only for our fellow Obamabots. When we chanted, "Yes we can," we meant all of us, Obama supporters, Hillary supporters, Republicans, independents. We had no doubt that everyone would soon see the light.

But first, we had to win. When we got back to campus, we turned a dorm room into a headquarters and traded call sheets for walk packets. Then we went from house to house, reminding supporters to vote. After finishing my packet, I cast my ballot, letting my eyes linger for just a moment on the filled-in bubble by Obama's name. Then I borrowed my roommate's car without permission and drove voters to the polls.

At the bar where we gathered to watch the returns, the mood at first was grim. Hillary was racing to an early lead. But then numbers rolled in from big cities—Hartford, Bristol, New Haven—and it became clear which way the night was headed. In the end, just fourteen thousand votes separated the two candidates. Still, that slim margin was enough. We had won. Whooping and hollering back to campus headquarters, our crew of college students passed a pair of panhandlers in the street. They looked at us. We looked at them.

"Barack Obama?"

"Barack Obama!"

Suddenly we were locked in a group hug, squealing as we leapt up and down. In an instant, our candidate's victory had bridged whatever gulf lay between us. I had every reason to believe the entire world had been no less fundamentally transformed.

It was Erika, a junior I was sort-of-kind-of dating, who brought me back to earth. I knew she was undecided on the morning of Election Day, but I wrote this off as a side effect of her major in philosophy. I was confident that, when the time came, she would make the right choice.

Instead, she voted for Hillary. She liked Obama, she explained, she just wanted to support a woman. In theory I was fine with her decision. In practice, it was as if we were at dinner and she had casually ordered human flesh. I tried to help Erika realize just how terribly she had chosen. For some reason, this didn't cause her to reexamine her life in the way I hoped. We sort-of-kind-of broke up a few days later.

By that time, though, Obama was on a roll. Nebraska. Maine. Maryland. Wisconsin. There were eleven contests between the Connecticut primary and February 19. We won them all. By the end of the month, Obama's lead in delegates was undeniable. His coalition of African Americans, well-to-do whites, and young people—an alliance so unlikely just a few months earlier—was now poised to make him the Democratic nominee.

With victory in the air, there was no time to mourn the end of a relationship.

Instead, I doubled down on hope. "There's this girl," I gushed. "I've never had the guts to ask her out, and she's never indicated she wants me to! Can you believe how much we have in common?"

Her name was Amy. She was spontaneous, rebellious, into both Batman and computer science. Clearly out of my league. But in moments of doubt, my candidate's words rang through my head. "When we have faced down impossible odds, when we've been told we're not ready or that we shouldn't try or that we can't, generations of Americans have responded with a simple creed that sums up the spirit of a people." The phrase had lodged itself in my head like a mantra, or a tumor.

Yes we can. Yes we can. Yes we can.

Throughout February, however, Amy politely hinted I could not. I tried every offer in my arsenal: home-cooked quesadillas, the final third of a bottle of Yellowtail, *The Sopranos* on DVD. She declined.

Then, just when it seemed hopeless, a window opened. Amy had caught the Obama bug. On March 4, the day of the Rhode Island primary, I asked her if she wanted to drop everything and go to Providence to knock on doors. To my surprise, she said yes. Before she could reconsider, I borrowed my roommate's car without permission and we took off down I-95.

From the moment we arrived, it was obvious that my candidate was not destined for victory. Rhode Island voters treated my suggestion to vote for Obama the same way Amy had treated my suggestion to meet up for a study break at 1 A.M. Yet even the most ardent fives had the sense something historic was happening. They disagreed with us, often strongly. Even so, they were happy to see young people on their front porches, working to change the country we all loved. One Hillary supporter offered lemon squares. Another asked me to come inside, worried I might freeze. Opening his door, a bearded man asked me a question I had yet to hear on the campaign trail.

"Are you Jewish?"

I told him I was, and he invited me into his modest white home to say Sabbath prayers with his family. I got the impression they would have consumed a heaping tray of bacon cheeseburgers before voting for my candidate. Still, they welcomed me with kindness in their eyes. This was the way politics was supposed to work. This was the way *life* was supposed to work.

Returning to the street to meet Amy, I felt more confident than ever that the universe was about to change. Who cared if we lost Rhode Island? America was on the brink of a new and shining chapter. We would be the ones to usher it in. I didn't feel disappointed. I felt elated. I felt powerful and giddy and unprecedented.

As Amy and I got back in the car, I knew she felt the same way, and I thought about leaning in to kiss her. But that seemed too ordinary. Too politics-as-usual. Too status quo. Before I had time for a second thought, or even a first one, I heard myself speak.

"Want to drive back to New Haven naked?"

My offer was totally unimaginable, and yet somehow within the realm of possibility. The Barack Obama of propositions. It took her just a moment to reply.

I don't remember how we got our clothes off while driving. All I know is that if they gave out merit badges for this sort of thing, we would have earned them. The city proved challenging, especially at speed bumps and red lights, but then we hit the highway and were free. I pulled into the left lane, zooming past drivers who gaped when they glimpsed the passenger side.

Was this a date? Was it a misdemeanor? Whatever it was, we didn't want it to end. We glanced at each other and laughed uncontrollably. We tensed nervously with every cop car we saw. We decided to break into an aquarium—trust me, it made sense at the time—and when that failed we pulled back onto the highway and continued our naked ride.

They said this day would never come, I thought.

I wish I could say my defining moment in politics was the caucus-night address that introduced me to Barack Obama, or the February rally when I first saw him in person. Or maybe the historic "race speech" six weeks later, long after whatever existed between Amy and me had sparked, sputtered, and flamed out.

But that would be a lie. The moment that changed my life came on a busy interstate, my bare cheeks pressed into the seat of my roommate's borrowed Nissan. Just a few months earlier, cynicism had flourished. But now? Now we were building a better world, a world where strangers invited you into their homes to break bread, and girls you had a crush on took their clothes off for no reason. We weren't just fighting for change. We *were* change. My foot was on the gas. My skin tingled with anticipation.

I was twenty-one years old, in Barack Obama's America. Anything was possible.

2

HOW TO NOT LAND A WHITE HOUSE JOB

DO NOT CLEAN THE OFFICE. OBAMA BOYS ARE THE JANITORS!!

Janice Maier, the chair of the Wayne County Democratic Party, didn't like me. In fairness, she didn't like Obama either. At more than eighty years old, a proud Hillary supporter, she viewed her candidate's young rival as an upstart. By extension I was a fetus, and an annoying fetus at that. It didn't matter that the primary campaign had ended a month ago. Janice still viewed my arrival with the frosty disdain of a Roman forced to welcome a Visigoth to the baths.

She couldn't bully me physically. At approximately four foot ten, she possessed both the stoop and jowls of a cartoon witch. But what she lacked in size she made up for in fiery, persistent hate. Hence the all-caps sign about janitors. Late one night I left our one-room campaign headquarters, and when I returned the next morning I found the note pasted against a wall. Only two people had an office key. I was one of them. The other was Janice Maier.

This was not the kind of challenge I had anticipated upon moving to Ohio. In May, the Democratic primary and my academic career ended with simultaneous whimpers. Two weeks later, I pointed my car toward a swing state. Obama for America had offered me something called an "organizing fellowship," a polite term for "indentured servitude." I worked unpaid, sixteen-hour days. In exchange, someone in headquarters found me a spare room in a stranger's house that smelled overwhelmingly of wet dog. I couldn't believe my luck. Each morning I popped out of bed, brushed dog hair from my button-down shirt, and happily harassed the good people of Canton, Ohio.

"Excuse me, are you registered?" I would ask.

"Nah, I don't vote."

"Terrific! Let's sign you up!"

When I wasn't registering voters, I was calling Obama supporters and inviting myself to their homes. The campaign dubbed the ensuing meetings "one-on-ones." If a one-on-one went well, the new volunteer would host a house party and play a motivational DVD for their friends. If it went really well, the volunteer might make an extravagant, heartfelt gesture, like bringing lasagna to the office for the staff.

"Remember," said Steph Speirs, one of our trainers in Columbus. "They come for Barack. But they stay for you."

I knew Steph from college. She graduated only a year before I did. But I soon learned that her youth was par for the course. Field organizers live in a state of unbridled meritocracy. There isn't time to defer blindly to elders or treat an org chart as sacred text. If your numbers are good—if you recruit lots of volunteers and contact lots of voters—you move up in the world. If not, you don't. Steph had hit the campaign trail only twelve months earlier. Now, she was in charge of Ohio's entire southwest quadrant. At our statewide training, she shared the mottos that helped her get ahead.

Don't try to win every vote. Just get to fifty plus one.

Your phone is your most powerful weapon.

Think with your head, be driven by your heart.

That last one sounds all kumbaya, but really wasn't. Meeting my weekly goals required a willingness—perhaps a slightly-too-willingness—to think of voters as numbers instead of humans. "There's a line of unregistereds outside the plasma bank!" I'd announce to my colleagues. "The best part is, they're just standing around!"

But I also discovered an unexpected warmth. One of my first one-on-ones was with Brenda, a middle-aged woman in a flowing blouse. As we parted ways, she looked at me nervously.

"Do you really think we can win?"

Ordinarily, I caveat almost everything. But in that campaign office, some long-dormant part of me stepped forward.

"Of course we can win! We've got Brenda on the team!"

Where the hell did that come from?

After a less-than-illustrious performance as a student—my four years of college were defined primarily by off-brand vodka and grade inflation—I worried my mediocre streak might continue. But in the field, something clicked. Before long I was collecting piles of voter registration cards. I was eating lasagna three meals a day. After a few weeks of indentured servitude, the campaign took notice, promoting me from unpaid fellow to underpaid organizer. Then they assigned me to open an office in Wooster, a small town in the heart of Northeast Ohio's dairy country.

Wayne County was what organizers called "tough turf," and not just because Janice Maier despised me. In 2004, Bush won the area with 60 percent of the vote. The remaining 40 percent was concentrated on the local college campus and in the poorer parts of town. Everywhere else, Democrats were political closet cases, afraid to openly declare their love for universal health care or a higher minimum wage.

This made our campaign a kind of coming-out party. Inspired by one of Obama's speeches, walk-ins would shyly open the door to our office, only to find friends and neighbors already there. Our little headquarters lay just off the town square, and it soon buzzed with the men and women I referred to, in campaign shorthand, as "vols." Ellen the silver-haired chain smoker. Sylvia the town busybody. Ross the unrepentant socialist. Beth the owner of a thriving mushroom farm. "You should be a used car salesman," they told me. Apparently, they meant this as a compliment.

I understood their point. Recruiting new vols required a certain lack of shame. But our organization was built on more than persistent begging and well-timed guilt trips. My greatest strength as an organizer was a monklike devotion to my cause. Between June and November, I drank exactly two beers. I was almost entirely celibate, almost entirely by choice. To avoid the distraction of national news, I downloaded porn-blocking software and reconfigured it to bar me from CNN.

If I had been selling cars instead of a candidate, I never could have been so single-minded. But each day in Ohio reminded me I was part of something big. Lisa had been laid off from three different manufacturing jobs in four years. When she stopped by in June, looking for a yard sign, she told me she was too scared to volunteer. By November she was a neighborhood team leader, with an entire ward of Wooster at her command.

Then there was Wendy. A slipped disk in her spine made both standing and sitting extraordinarily painful, so she called voters while pacing in agonizing laps around the town square. I took pride in pushing vols to the limit, but this time, even I begged her to stop.

"You don't have to keep doing this."

"Yes, I do. If I don't make these calls, I won't be able to get the health insurance I need to get better."

With so much on the line, who wouldn't work 120 hours a week?

Thanks to my porn-slash-news blocker, I didn't fully appreciate that as our organization was growing, the economy was falling apart. On September 15, Lehman Brothers collapsed, but I was too busy planning a Joe Biden rally to care. My vols were less lucky. With lives outside the campaign, they couldn't afford my blissful ignorance. In between phone calls, they had once complained about Bush. Now they discussed their 401(k)s. Back in Canton, coworkers canvased a street only to find that every house had been foreclosed on.

With the looming threat of a second Great Depression, I feared our supporters would be too busy worrying about their own futures to spend time with the campaign. Instead, the opposite occurred. More than ever, volunteers could see the connection between the national life and their own. They began acting with what Obama called "the fierce urgency of now." As they picked up the pace, so did I. By October, I had discarded Steph's mottos in favor of a new, more personal one.

Tiredness is just a feeling.

And then came Election Day, which was oddly, almost eerily calm. Hours before polls opened, I plopped onto a couch in a volunteer's living room and waited for a crisis. When none came, I found an empty two-liter bottle of juice and an old Nerf football. I spent one of the most important days in American history seeing how many bullseyes I could hit.

Thwap. Thwap. Thud-crash-profanity-sound-of-feet-across-carpet-to-check-if-lamp-is-broken-sigh-of-relief. Thwap.

Outside my boiler room, the Wayne County ground game was in high gear. But it was no longer run by paid operatives. Teams of trained local volunteers directed it themselves. They were students, professors, stay-at-home moms, karate teachers, agricultural researchers, retirees. And now, they were organizers, too.

That night, when the networks called Ohio and then the country for Obama, we couldn't believe what he had accomplished. More than that, we couldn't believe what *we* had accomplished. True, the economy was still in crisis. True, the Iraq War still raged. True, Janice Maier would soon steal the laser printer I had purchased at OfficeMax, refusing to release it until I scoured her folding table with a sponge.

"You can't just hold people's stuff hostage!" I would say. For the first time in months, her lip would curl into a smile.

"That's the way we do things in Wayne County, buster."

Relegated at last to janitor, I would have no choice but to scrub. But that was in the future. Right now, our new president-elect was addressing the country. My vols were weeping with joy. The arc of the moral universe was bending toward justice at last.

THERE IS NOTHING QUITE LIKE ELECTING A PRESIDENT AND THEN moving back in with your parents. I can't say I recommend it. On the campaign trail, I was treated like an all-star. In my childhood bedroom, I spent my nights staring at faded summer camp certificates declaring me Most Improved.

It wasn't just inside my mom and dad's apartment that I felt demoted. New York City's stream of small humiliations reminded me that I had become a kind of human plankton.

Brooklyn lumbersexuals cut in front of me at food trucks. Briefcase-wielding bankers sideswiped me on their way to work. I had no job. I had no career path.

All I had, really, was an envelope. It arrived in early January, oozing exclusivity and class.

The Presidential Inaugural Committee requests the honor of your presence.

Thick, expensive card stock. Letters embellished with loops and tails. The year written out in full: *two thousand and nine.* I had never possessed anything so fancy.

Along with the finely printed note came a pair of invitations, one to the official ceremony at the Capitol, the other for an inaugural ball. I guarded my tickets as fiercely as Charlie before his chocolate-factory tour. And for good reason: with the swearing-in just a few weeks away, scalpers were charging up to twenty thousand dollars for an inauguration package like mine. I could have made a fortune on Craigslist. Yet I would sooner have sold a kidney. While the rich and well connected scrounged for access, a twenty-two-year-old with a Koosh basketball hoop on his door had the most coveted tickets in town. What better proof that a brighter future was at hand?

I didn't go to D.C. hoping to land a job with Obama, any more than a Phish fan goes to a Phish concert hoping to land a job with Phish. Still, not long after receiving my envelope, I decided it was time to move. With five days left in George W. Bush's second term, I zipped my precious tickets into my suitcase. Then I added five button-down shirts, two pairs of khakis, and the tuxedo I wore to prom. I still had no career path. I still had no plan. But hope and change were on their way to the nation's capital. In that case, so was I.

HOURS BEFORE INAUGURATION DAY DAWNED, I MADE MY WAY TO downtown D.C. It was brutally, freakishly cold. As I followed the instructions on my purple ticket, walking toward a designated meeting area near the Capitol, I shivered despite my puffy coat.

To make matters worse, when I reached the rendezvous point, it became obvious police hadn't reserved enough space. Thousands of freezing former organizers were being crowded into an enclosure no larger than a basketball court. By 7 A.M., chemically speaking, we had become a solid.

Yet penned in with my fellow campaign veterans, I didn't feel self-pity. I felt pride. While so-called VIPs might demand special treatment, the new Washington had no room for outsize egos. We were all created equal. Now we stood an equal chance of being crushed to death. How much more egalitarian can you get?

I was laying track on this high-minded train of thought when a whisper began to ripple through the crowd.

Jesse Jackson. Jesse Jackson? It's Jesse Jackson!

I didn't think there was space for a squirrel to move among us, much less a broad-shouldered mountain of a man. But here he was, a giant of the Civil Rights era, all six foot three of him, coming toward me in an enormous black greatcoat. To my amazement the crowd parted, creating a kind of human corridor, and Jesse Jackson did not walk through it. He strode. His expression was less imperious than magnanimous, as though he were posing for a portrait he planned to give away. Expensive wool swished against the nylon shell of my jacket. I stood awestruck, too stunned to speak.

That's when I realized something: the living legend wasn't the only one making progress. A sneaky little man had inserted himself into the reverend's wake. He was middle aged, with a white, bushy mustache, a hat he appeared to have borrowed from Indiana Jones, and a firm grip on the back of Jesse Jackson's overcoat. While the rest of us pressed against each other, he zipped along like one of those fish that clean the bellies of sharks. Our eyes met as he passed me. With his free hand, he gave me a wave.

"Hey hey!" he cried. Before I could stammer a reply, the corridor of humanity closed behind him. The sneaky little man disappeared.

The rest of us resumed waiting. A half hour later, we were told to line up in a tunnel beneath the National Mall. If we had been at the post office, this is the point when someone would have thrown a chair through a window. But on Inauguration Day there were no complaints. We formed a queue of shivering purple-ticket holders several blocks long. It was only after another hour passed, and we still hadn't budged, that a new whisper began worming down the line.

They closed the gate. They closed the gate? They closed the gate!

It was true. Not long after Reverend Jackson passed us, a mishap occurred at the purple entrance. Secret Service shut it down. My only hope of seeing the inauguration now was to find a TV, and fast. I ducked out of line, chose a direction, and ran until I reached a bar.

In the tunnel, my neighbors were intoxicated by the historic nature of the moment. In the bar, my neighbors were intoxicated because they'd been drinking since 6 A.M. The place combined the macho chaos of a frat house with the undersexed intensity of a debate team. Each time a female legislator appeared on C-SPAN, catcalls filled the room.

"Nancy Pelosi's hot!"

"Dianne Feinstein's hot!"

For some of my fellow ticket holders, this was the precise moment when hope calcified into cynicism. By the next morning they even had a name for the tragedy: The Purple Tunnel of

Doom. And while I didn't share their sense of full-blown catastrophe, even I had to admit that something had changed. A few nights later, I was waiting to enter the Staff Inaugural Ball when someone blatantly cut in line. It was Miranda, a fellow Ohio organizer.

"Everyone does it," she informed me.

Our movement has no place for line cutters, I thought. But part of me wondered if she and a tiny man with a white mustache knew something I did not.

I FIGURED FINDING MY PLACE IN OBAMA'S WASHINGTON WOULD BE quick and straightforward, like learning your house at Hogwarts. But just to be on the safe side, I found an internship with a crisis communications firm while I waited for the sorting to begin. This was a boutique agency, the kind that boasted former lawmakers on the board. It was also the kind of place that makes you sign an agreement prohibiting future use of its real name.

"You'll love the Crisis Hut," my interviewer promised over the phone. She knew I came from the Obama campaign, and she assured me her own organization was no less devoted to hope and change.

"One of our clients is a company that manufactures cribs. When a crib collapsed and killed a baby, we helped them get a second chance!"

I accepted her offer, confident I would be a Crisis Hut employee for no more than a few weeks. I was a former organizer with outstanding numbers. How long could finding something take?

But Obama had hired more field staff than any campaign in history. There simply weren't enough jobs to go around. The Ohio campaign team held regular check-in conference calls. Listening to them was like reading logs from an Arctic expedition as it slowly realizes it's doomed.

January 21: Fired up! Ready to go!
February 9: Stay positive. Positions take a long time to fill.
March 16: Ask yourself: Am I setting my sights too high?
April 3: Remember, it's okay not to get a job.

It was during these months that I realized the phrase *Before Iowa* was being used in Washington the way *Before Christ* was once used in Galilee. It seemed only fair that those who joined the campaign BI, like Steph Speirs, got first pick of jobs. But other hiring decisions were harder to swallow. Miranda, the inaugural ball line cutter, joined the campaign after I did, yet she landed a coveted administration post just weeks after arriving in D.C.

By that time I was frustrated, but not surprised. Even the new Washington, it seemed, rewarded a particular set of skills. A talent for flattery. An unshakable sense of entitlement. A sense of confidence out of proportion to one's achievements. I don't mean to imply that 2009 was an endless series of *House of Cards* sex-murders in a desperate climb to the top. All else being equal, however, a slight inclination toward sex-murder didn't hurt.

Nor was I above playing the game. A friend of a friend was twenty-six years old, a bona fide professional working in health care policy, and I was thrilled when he agreed to meet me for coffee. For thirty minutes, I tried desperately to soak in his wisdom. He saved the best for last.

"Remember," he proclaimed, "networking is only bullshit if *you're* bullshit." Then he headed for the door, tossed his paper cup in the recycling, and never replied to my e-mails again.

After the Obama campaign, working at the Crisis Hut felt like joining the marketing department of Soylent Green. Our clients included a Wall Street bank that helped wreck the economy, a coal company that considered worker safety a form of socialism, and a gold magnate whose mining process involved small lakes full of cyanide. Facebook updates from former coworkers were like postcards from a country where my visa had been denied. "We got the stimulus!" posted one former Ohio colleague, now a White House staffer. I was happy to see President Obama's economic package pass Congress. Still, I winced. This was a new kind of *we*. I wasn't included. My movement was moving on.

Things only got worse. By March, friends who had yet to land a job began slinking out of D.C. like baboons rejected by the troop. This was around the time I informed Gertrude, my Crisis Hut supervisor, that our mining magnate's profits were threatened by jewelers using reclaimed gold.

"Stupid environmentalists," she replied, in a tone that struck me as oddly personal. "Always trying to be green, am I right?"

Looking back, I had reached what campaign organizers refer to as a "choice point." On one hand, I could ditch the Crisis Hut, abandon the job search, and return to a life of Manhattan planktonism. On the other hand I could stay, making my peace with disaster-prone coal companies and cyanide-filled lakes. But both paths felt like surrender. I wasn't ready to choose. Instead, with my work life spiraling downward and my dreams of changing Washington unfulfilled, I did what every D.C. intern dreams of doing. I went rogue.

It started with the dress code. I began wearing shirts I'd bought during a study-abroad program in China, a collection whose dominant theme was "polyester." Next I brought my laptop from home, abandoned my cubicle, and set up shop in the break room. If the economy hadn't been shedding hundreds of thousands of jobs each month, perhaps my coworkers would have complained. But preoccupied by weightier concerns, they let me go about my business.

More and more often, *my business* was a euphemism for *playing Minesweeper.* It was one of two free games that shipped with my laptop, and I began by clicking through the easy grids, ten by ten. Soon, however, I was playing six to eight hours a day, and had advanced all the way to a hundred by a hundred. When even this failed to attract notice, I came up with my boldest act of rebellion yet: answering questions exclusively in analogies to the game of Minesweeper I was currently playing.

"David, did you finish that report on infrastructure investment?" Gertrude might ask.

"Almost," I'd reply. "But you know how when you're playing Minesweeper? And you're on the highest difficulty setting? And you only have one mine left but it's taking you a long time to decide where to click? It's kind of like that." I'd drum my fingers on the keyboard, hard at imaginary work.

"I might not be done for a while."

I congratulated myself on my cunning. But I was an intern Icarus, flying too close to the sun. Late one morning, straggling to work and draped in synthetic fabric, I heard a voice behind me. It was Bill, Gertrude's angular, shark-faced boss.

"My office. Now."

Less than three years later, I'd meet with White House interns hungry for advice. How had I managed to land a dream job? I would play my part, furrowing my brow and mumbling something about doing what you love. But in those moments, what I really thought about was the way my shoes sank into the beige carpet as I walked to Bill's corner office. From a burnished wood picture frame, his wife and son stared at me accusingly. Once seated at his desk, my boss's boss leaned backward, shark eyes flashing with a mix of anger and bewilderment. In the break room, I had felt defiant and alive. Now I just felt childish.

I learned something about myself that day: I am not above begging for forgiveness.

I was lucky to receive it. Most rogue interns never get a second chance. And here it's worth mentioning that I benefited from what was known in 2009 as being fortunate, and is now more commonly called privilege. It's not like I flashed an Ivy League gang sign and was handed a career. If I had stood on a street corner yelling, "I'm white and male, and the world owes me something!" it's unlikely doors would have opened. What I did receive, however, was a string of conveniences, do-overs, and encouragements. My parents could help me pay rent for a few months out of school. I went to a university lousy with successful D.C. alumni. No less significantly, I avoided the barriers that would have loomed had I belonged to a different gender or race.

Put another way, I had access to a network whether I was bullshit or not. A friend's older brother worked as a speechwriter for John Kerry. When my Crisis Hut term expired, he helped me find an internship at West Wing Writers, a firm founded by former speechwriters for Bill Clinton and Al Gore. In the summer of 2009, my new bosses upgraded me to full-time employee.

Without meaning to, I had stumbled upon the chance to learn a skill. The firm's partners were four of the best writers in Washington, and each taught me something different. Vinca LaFleur helped me understand the benefits of subtle but well-timed alliteration. Paul Orzulak showed me how to coax speakers into revealing the main idea they hope to express. From Jeff Shesol, I learned that while speechwriting is as much art as craft, and no two sets of remarks are alike, there's a reason most speechwriters punctuate long, flowy sentences with short, punchy ones. It works.

The firm's fourth partner, Jeff Nussbaum, had carved out a niche writing jokes for public figures. It was he who taught me about the delicate balance all public-sector humorists hope to strike. Writing something funny for a politician, I learned, is like designing something stunning for Marlon Brando past his prime. The qualifier is everything.

At first I didn't understand this. In June, President Obama's speechwriters asked Jeff to pitch jokes for an upcoming appearance at the Radio and Television Correspondents' Dinner. I sent him a few ideas, including one about the president and First Lady's recent trip to see a Broadway show:

"My critics are upset it cost taxpayer dollars to fly me and Michelle to New York for date night. But let me be clear. That wasn't spending. It was stimulus."

Unsurprisingly, my line about stimulating America's first couple didn't make it into the script. But others did. The morning after the speech, I watched on YouTube as President Obama turned to NBC reporter Chuck Todd.

"Chuck embodies the best of both worlds: he has the rapid-fire style of a television correspondent, and the facial hair of a radio correspondent."

That was my joke! I grabbed the scroll bar and watched again. The line wasn't genius. The applause was largely polite. Still, I was dumbfounded. A thought entered my brain, and then, just a few days later, exited the mouth of the president of the United States. This was magic.

Still, even then, I had no illusions of becoming a presidential speechwriter. When friends asked if I hoped to work in the White House, I told them Obama had more than enough writers already. I meant it.

Besides, working for well-heeled clients came with its own set of perks. One of our clients, a bona fide billionaire, owned a pro basketball team. A few months after starting my new job, I was invited to a game. As I took my plush seat in a luxury box, I marveled at the bounty before me. Free beer. An amazing view. Unlimited hot dogs. The perfect afternoon.

Then the subject turned to Obama's proposed health care

law, and suddenly everything changed. My host's eyes took on a wistful, faraway quality.

"I'll tell you," the billionaire sighed, looking out from the owner's box. "It just doesn't pay to be rich anymore."

In fairness, our client was not the only one who had come slightly unglued over Obama's policies. On April 15, 2009, I was walking through downtown Washington when I came across a crowd of people railing against government and waving tricornered hats. It was hard to take them seriously while they were wearing Halloween costumes. Imagine watching someone dressed as a sexy kitten endorse a Medicare privatization scheme. Still, I shouldn't have dismissed them. On television and in person, the conservative colonists multiplied with alarming speed.

Some of the leaders of the backlash said their name was an acronym for "Taxed Enough Already." Maybe this was true at first. But the Tea Party was soon infused with paranoia that had nothing to do with taxes. While the ugliness caught Washington observers by surprise, anyone who had spent time in a battleground state recognized it instantly. Back in Ohio, volunteers had been told to check boxes corresponding to a voter's most important issue: economy, environment, health care. But what box were you supposed to check when a voter's concern was that Obama was a secret Muslim? Or a terrorist? Or a communist? Or the actual, literal Antichrist? How could you convince a voter whose pastor told them your candidate would bring about the biblical end of days?

Other people were just plain racist. Outside an unemployment center in Canton, a skinny white man with stringy hair and a ratty T-shirt told me he would never, ever support my candidate. When I asked why, he took two fingers and tapped them against the veiny underside of his forearm. At first I didn't understand.

"You won't vote for Obama because you're a heroin addict?"

It took me at least ten seconds to realize he was gesturing to the color of his skin.

This sort of thing was not part of every conversation. It would be wrong to say racism was the only thing eroding Obama's support. But it would be equally wrong to deny its existence. Better to say that bigotry was like one of the Beach Boys who wasn't Brian Wilson. The lyrics would have been the same without those unmistakable *ooohs* and *ahhs,* but it would have been a whole different sound.

For the first few months of the campaign, this paranoid, prejudiced stew simmered in the background. Then John McCain picked Sarah Palin as his vice president, and everything blew up. As a speechwriter, I learned to use the phrase *permission structure* to describe the conditions that allow a choice to be made. If you're counting calories, for example, "cheat day" might be the permission structure for a plate of deep-fried chicken wings followed by several slices of chocolate cake.

Sarah Palin was a permission structure unto herself. With the mannerisms of June Cleaver and the worldview of Joe McCarthy, she gladly trumpeted what the Bush administration merely implied.

"We believe that the best of America is in these small towns that we get to visit, and in these wonderful little pockets of what I call the real America."

"He's palling around with terrorists."

"I am just so fearful that this is not a man who sees America the way you and I see America."

Suddenly, every day was cheat day. In the countryside, volunteers reported that Obama yard signs were being used for target practice. Mailboxes were bashed off their posts in the middle of the night. One day I ducked into a Panera at the shopping center. When I returned to the parking lot, I found an elderly white man in a sport coat verbally assaulting the bumper sticker on my car. Perhaps he was protesting the idea of a black president. More likely, skin color was just one of many things that made Obama seem so wrong. His big-city background. His youth. His Ivy League degree. His name. The fact that Ludacris was on his iPod and Pat Boone was not. There was an endless buffet of otherness to choose from.

At first I imagined that, in the same way that only the winning candidate gets to live in the White House, only the winning philosophy gets to come to D.C. During Obama's first hundred days, that appeared to be the case. Congress expanded health insurance for low-income children, invested in clean energy and infrastructure, and helped protect equal pay. The president created a task force to save the auto industry, expanded women's health funding, and reversed George W. Bush's ban on stem cell research.

But my movement wasn't the only one with big ambitions. On September 12, 2009, more than seventy-five thousand Tea Partiers marched on Washington, a pale sea dotted with pastel lawn chairs and yellow *Don't Tread on Me* flags.

Once again, Sarah Palin gave voice to the crowd. On August 7, she had claimed the president's proposed health care law included "death panels." This was categorically false. It would have been equally honest to decry Obama's plan to lay eggs inside your brain. But the Tea Partiers believed it anyway. As tricornered hats advanced toward the National Mall, the phrase was on everyone's mind.

The dishonesty made me furious, but as a budding speechwriter, I had to give Sarah Palin credit for a well-crafted slogan. *Death panels* had captured the Right's imagination no less thoroughly than *Yes we can* had captured the Left. The middle ground was shrinking. You had to pick a side.

Team Yes We Can continued notching victories. On

March 23, 2010, President Obama's sweeping health care overhaul finally passed. I thought of Wendy, my vol from Ohio. My eyes filled with tears. But Team Death Panels was gaining strength. For Republicans in Congress, voting on one of Obama's laws had become like critiquing one of Hitler's paintings. You were required to hate it, whether you liked it or not.

As the November elections drew closer, an anti-Obama wave began to build. My bosses at the speechwriting firm gave me five weeks off to try and stem the tide. From a tiny cubicle in Democratic National Committee headquarters, I resumed recruiting vols and calling voters, but this time there would be no Election Night relief. We lost every single congressional race we hoped to win. With newfound control over the House of Representatives, Republicans now held veto power over any law the president hoped to pass.

I hadn't stopped believing in Obama. But the idea of a new Washington had never seemed so absurd. It was time to leave the nation's capital. I sublet my room to a producer for NPR music, sent invites for a going-away party, and began looking for an apartment in Chicago. Obama's reelection campaign was setting up its headquarters there. I would knock on the door, show them my numbers from Ohio, and unpack boxes until somebody hired me.

But there was something I didn't fully appreciate: hacking away blindly at the vines and shrubbery of life, a career path had emerged. In an act of remarkable generosity, my bosses sent my writing samples to David Axelrod, President Obama's messaging guru, and his chief speechwriter, Jon Favreau. Jon asked for my résumé. Thanks to my 2010 campaign work, we had mutual friends who could vouch for me. One of these friends introduced me to Tyler Lechtenberg, a speechwriter for the First Lady, who quietly suggested I might not want to leave town. Without realizing it, I had become not bullshit.

A few days later, Jon and I met for coffee. His deputy was leaving, he told me. I could add my name to the long list eager to join his team. But there was another option. Valerie Jarrett, the president's senior advisor, had spent months looking for a speechwriter without success. If I was interested, I would become the sole candidate for the job.

It was, in other words, a choice point. Would I put my faith in meritocracy? Or would I seize the chance to cut in line? Two years earlier, I would have eagerly applied to be a presidential speechwriter, confident the best person would win. Not anymore. I told Jon I wanted the job with Ms. Jarrett. Then I double-crossed the NPR producer, took back my room, and made sure I owned a suit, shirt, and tie.

You know how when you're playing Minesweeper? And at first you just can't get the hang of it, no matter how hard you try? But then everything begins making sense, even on the highest difficulty setting, and the goal you never even admitted having is suddenly within your reach?

It was kind of like that.

A few days later, feeling slightly stuffy in my suit, I made my way to a small kiosk a few blocks from my firm's office. "Can I help you?" asked a Secret Service agent, from behind his pane of bulletproof glass.

"Yes, you can," I told him. And then, just because I could, I added something.

"I have an appointment at the White House."

3

CLEARED TO WORK

The intern escorting me to my interview had the nervous formality of a bar mitzvah boy. Mike Strautmanis, Valerie Jarrett's chief of staff, was clearly more comfortable in his skin. Above his office fireplace, instead of a portrait or plaque, he kept a pair of signed Yao Ming sneakers. Next to his conference table he had installed one of those putting-practice machines that spits the ball back when you make a shot. A different kind of high-ranking official might have held it against me when I mispronounced his name, rhyming the first syllable with *trout* instead of *trot*. Straut let it slide.

I had spent days rehearsing my answers to interview questions. Favorite Obama speech. Greatest strength. Greatest weakness (that was sneakily also my greatest strength). But Straut didn't ask me any of that. Instead, he rose from his chair, towering over me as he headed for the door.

"Tell you what," he said. "I'm leaving for an hour. Why don't you write a two-page speech for a breakfast roundtable with CEOs. Then I'll take a look."

This was madness. At my firm, I'd have at least a week for a two-pager. I'd also be given guidance on what to write. But with the clock ticking, there was no time to complain. I dove in, a white-collar MacGyver defusing a rhetorical bomb.

All right, David, think. Quotes from the president. Random facts about the economy. Praise for American innovation. It just might work.

I finished my draft mere seconds before Straut returned. While he read through it, the bar mitzvah boy intern brought me to a greasy, ground-floor cafeteria known as Ike's. I ordered a ham and cheese on a thick sub roll, and ate it with trembling fingers.

At least I could take comfort in being finished. In the worst-case scenario, I could always say I had written a speech in the White House. In the best-case scenario, I would return a few days later for a follow-up with Ms. Jarrett herself.

"Tell you what," said Straut, when I got back to his office. "Let's see if she's free right now."

"Sounds great!" I replied.

But I didn't mean it. As far as I was concerned, a meeting with Valerie Jarrett—Senior Advisor to the President, Head of the Office of Public Engagement and Intergovernmental Affairs—required the sort of preparation usually accompanied by a montage. Valerie was one of the most influential people in the White House. This meant she was one of the most influential people in the world. Even her lengthy Wikipedia page made her appear both glamorous and terrifying, a cross between Anna Wintour and the Sphinx.

Which is why I was surprised when our meeting was quick, painless, and disarmingly low key. Valerie was warm and friendly, albeit in an official, don't-forget-that-I-could-squash-you sort of way. She asked about my experience. I told her I was excited about the opportunity. Every so often she nodded thoughtfully, and ten minutes later we were done. Only after I left her office, and Straut began asking me about salary, did I realize Jon Favreau hadn't been exaggerating. There really were no other candidates. The job was mine.

When strolling down memory lane, it's always tempting to spackle on a layer of retroactive dignity. *As I considered my good fortune, John F. Kennedy's words echoed through my heart. "Ask not what your country can do for you," I told myself, thanking God for the gift of freedom.*

Nonsense. I rushed back to my apartment, ripped off my suit, and jumped up and down in my underwear. I fist-pumped. I hollered the word *holy,* followed by every obscenity I knew. Then I immediately started calling people.

I SHOULD HAVE BEEN MORE CAUTIOUS. GETTING A JOB OFFER FROM the White House is like getting a marriage proposal from Tom Cruise: there's a lot of paperwork before the deal is sealed. The due diligence began before my interview was even over, when Straut asked if there was anything I hadn't told him yet.

"It's better if you let us know," he said. His tone was part friend, part hitman. "We've 'unhired' people before." With these words, the background check began.

First came the federal investigation. Friends and family were interrogated by the FBI, their stories cross-checked for inconsistencies and gaps. Next was a form called the SF-86, a 127-page Godzilla of questionnaires. I wrote down every address I'd ever lived at. I listed every job I'd ever held. Some questions were fairly standard: Had I been convicted of a crime? Had I been delinquent on my debts? But others reflected a curious faith in the power of capital letters, and the honesty of America's foes.

"Are you now or have you **EVER** been a member of an organization dedicated to terrorism?"

"Have you **EVER** knowingly engaged in activities designed to overthrow the U.S. government by force?"

Most of the SF-86 was nerve-racking without being truly scary. It's the feeling you get when you go through airport security and a tiny part of you wonders if you packed a bomb. For Democrats my age, however, there was one frightening exception: drugs. It used to be that any substance use whatsoever was an automatic deal breaker for federal jobs. That's no longer the case. But plenty of unwritten rules still apply, and in 2011 they were the subject of a churning rumor mill among the young people of Washington, D.C.

As long as it was just in college.

As long as it wasn't cocaine.

As long as it happened in Amsterdam.

As long as you didn't deal.

The system could be cruel. In 2009, a West Wing Writers associate named Tom was hired by a cabinet department. On his SF-86, he wrote that he had "regularly" used marijuana in college. This was a mistake. The way Tom explained it, a federal investigator consulted some dusty chart from the *Reefer Madness* era. Failing to find an official definition of *regularly,* he had substituted *habitually.* According to the chart, *habitually* meant *weekly.* According to the chart, *weekly* meant *addicted.*

If Tom had checked into rehab and kicked his nonexistent habit, he might have stood a chance. Instead, by decree of the United States government, he was both drug addled and unreformed. Application denied. I arrived at West Wing Writers just as Tom began appealing his case. He used dictation software, so for an entire week I could hear what sounded like a forced confession being recorded down the hall.

"On . . . March . . . twelfth . . . 1999 . . . I . . . hit . . . my . . . roommate's . . . bong."

Determined to avoid Tom's fate, I got specific. After some back-of-the-envelope math, I listed thirty instances of undergraduate marijuana use, plus one experience with mushrooms I made clear I hadn't enjoyed. Afterward, I proudly shared my numbers with a friend who worked for the National Security Council.

"Thirty?! You should've told them less than ten!"

While the FBI was making sure I wasn't a threat, White House lawyers were making sure I wasn't an embarrassment. I write these words during the early days of the Trump presidency, when rejecting unsavory applicants seems as quaint and old-timey as canning your own peaches. But in the olden days of a few years ago, the vetting process struck fear into our hearts.

Vetting wasn't entirely subjective. If you were like some of my Yale classmates, ducking out of photos at parties so you could one day run for office unblemished, you would probably pass vet. If you enjoyed tweeting about female anatomy or spinning records under the stage name "DJ White Power," you probably would not. If you were like me, however, you found yourself in purgatory. Lindsey, an associate with the White House Counsel's Office, called me for a friendly interrogation, but I had no idea who she was calling next. Were exes fair game? What about the old, deeply regrettable stand-up routines from my high school talent shows?

I was told my application had been prioritized, but even so it took forever. One week I was asked to confirm some personal information. The next week I was invited to the Interior Department to pee in a cup. Throughout March I was a contestant on a federally run reality show, bouncing between challenges, unsure whether I'd be eliminated or handed the final rose. Each day my heart pounded as I scanned my inbox. Had something gone wrong with my urine sample? Had the vetters learned about the month in college when I paired thrift-store blazers with Looney Tunes pajamas and was certain I had started a trend?

Finally, on March 30, 2011, I got an e-mail from Alex, Straut's assistant. I was twenty-four years old. Most of my proudest achievements in life still fell under the category of "overcomplicated pranks." But starting April 1, I was cleared to work in the White House.

MY FIRST WEEK AT WORK, I PUT A FULL CUP OF COFFEE THROUGH THE X-ray machine by the security scanner. It emerged a few seconds later, empty and on its side. "Not a lot of metal in that," sighed the Secret Service agent on duty, looking for a towel. "Just a thought, for next time." I could practically see him add me to his Do Not Save list.

I had not expected this sort of thing to happen. I guess I thought walking through the White House gates instantly makes you better, stronger, and more capable than before. It doesn't. While my job was more exciting, the rest of me remained fundamentally unimproved.

This was grossly unfair. Every day, my unchanged abilities were pitted against drastically heightened expectations. Just a few months earlier, for example, I had panicked at the idea of a weekly roommate check-in meeting to discuss the chore chart. Now, with Valerie set to speak on international women's issues, I was supposed to bring together a half dozen experts from the National Security Council staff.

What made this meeting particularly frightening was that it included Samantha Power, the president's top advisor on human rights. At forty-one years old, she already had a degree from Harvard Law, a Pulitzer Prize, and years of experience on the

White House senior staff. Far smarter people than I had been exposed as morons merely by her presence. When the big moment arrived, I nervously made my way to a spacious NSC office suite. Then I opened the door, certain one of the greatest foreign-policy minds of her generation was about to reduce me to intellectual rubble.

But Samantha Power was running behind schedule. This was a golden opportunity. If I could win over lower-ranking staffers, I'd have allies in the room when the president's advisor arrived. A half dozen of us pulled chairs into a circle. I did my best to turn on the charm. With a shock of satisfaction, I realized it was working. I was making small talk! I was fitting in! Everyone was smiling broadly when, too late, I realized I already knew the stony-faced policy savant sitting directly to my right.

"Funny story," said Richard, when his turn came to introduce himself. "David and I met years ago. I was actually his teaching assistant in college." At this, another staffer, one about twice my age, saw a chance for some lighthearted fun.

"So, was he a good student?"

I will never forgive Richard for answering honestly. "You know," he said, eyebrows arching like an *Us Weekly* photographer seated next to Lindsay Lohan on a flight, "he seemed kind of bright. But I always got the feeling he didn't want to be there. He never really applied himself."

At that exact moment, as six get-to-know-you smiles turned to frowns, Samantha Power walked in. I tried to apply myself. I really did. But it was hopeless. Each time I asked for details, or failed to grasp a point, I was met with a half dozen disappointed stares.

The disapproval of my former teaching assistant I could live with. The disapproval of my new boss was something I hoped to avoid. This was easier said than done, as Valerie Jarrett's superpower was the ability to ask the one question for which I had forgotten to prepare. Before taking my seat at the burnished wood conference table in her West Wing office, I'd research everything: the size of the crowd, the name of the introducer, the exact length of the remarks.

"So," she'd begin, "who's speaking after me?" I don't know how she did it. This happened every time.

It was moments like these—plus her habit of looking at you like a goldfish she appreciated but wouldn't be heartbroken to flush—that accounted for Valerie's fearsome reputation. "She expects her staff to work as hard as she does" was the line staffers often used. In Washington, this is typically a euphemism for "harasses interns" or "once threw a stapler at the scheduler's head," but in Valerie's case it was true. In 1991, she had interviewed a young woman named Michelle Robinson for a job in Chicago's city hall. Not long after, she was introduced to Michelle's fiancé, a young lawyer named Barack.

The Obamas had been family ever since. More than nearly anyone, Valerie knew how improbable their journey had been. She was determined not to squander a single moment. She demanded the same determination from everyone she employed.

Valerie never told me this directly, but I also came to believe she saw herself as a custodian of the president's conscience. As senior advisor, she served not just the middle-aged politician in the Oval Office, but the young idealist she had met two decades before. Ending "Don't Ask, Don't Tell." Fighting inner-city poverty. Opening a gender-neutral White House bathroom. When something fell under the category of "not necessarily great politics, but the right thing to do," it was often Valerie leading the charge.

This combination of intense loyalty and unabashed progressivism made my new boss a favorite target. *The mouthpiece of the real POTUS, Comrade Valerie Jarrettikov!* It wasn't just Internet trolls who despised her. The mere mention of her name could derange even the most polished member of the GOP elite. "She seems to have her tentacles into every issue and every topic," declared Utah congressman Jason Chaffetz, as though Ursula from *The Little Mermaid* had taken up residence in the West Wing.

But if there was any theft of voices, or abuse of poor unfortunate souls, I never saw it. Valerie's portfolio was broad because all senior advisors' portfolios are broad. Involving herself in a wide variety of issues—as both advocate and enforcer—was the essence of her job. And in April 2011, my first month at the White House, those roles had never been more critical. The unemployment rate was nearly 9 percent. Household incomes were going down instead of up. With Election Day less than two years away, there was zero margin for error, and no room for careless mistakes.

This was especially true for the youngest members of the staff. The small inner circle responsible for President Obama's rise was not about to let someone born after the first *Ghostbusters* be responsible for his fall. At each semester's orientation, Straut lulled new interns into a false sense of security with talk of friendships and personal growth. Then, without warning, he lowered his voice.

"Let me be absolutely clear," he said. Too late, the interns realized their teddy bear had become a grizzly. "If you do something stupid and end up on the front page of the *Post,* you will be out of here immediately. I will not ask for your side of the story. I will not give you a second chance. I will not feel sorry for you. In fact, I will never think about you ever again."

You could practically hear the sound of hearts sliding into throats.

I was the only speechwriter at these meetings. Most of Valerie's employees in the Office of Public Engagement (OPE, for short) were "liaisons," professional extroverts each responsible for a slice of the Obama coalition. Young people, African

Americans, Hispanics, Jews, state legislators, enviromentalists: an eager young person was assigned to them all. Where other offices thrived on gallows humor, our corner of the White House buzzed with earnest optimism.

"*We put the OPE in HOPE!*" my coworkers wrote in e-mails, without any trace of irony whatsoever.

I also attended meetings of the POTUS speechwriting team, where, to my relief, sarcasm thrived. Still, I wondered if Jon Favreau would welcome me in. Thanks to his youth, and the quality of his work, President Obama's twenty-nine-year-old chief speechwriter had become something of a local celebrity. He dated Hollywood actresses. Gawker posted pictures of him playing beer pong without his shirt. I was no good at beer pong. I couldn't begin to imagine a scenario in which gossip blogs clamored for my shirtless pics. I worried that Jon and I might not get along.

What I hadn't realized was that, above all else, Favs was a prodigy. Speechwriters, even great ones, tend to lead either from the head or heart. I was a head-first writer, connecting logical dots and only later adding emotions. Heart-first people went the other way around. Favs was the only true switch-hitter I ever met. His writing was both lyrical and well organized, arcing between timeless values and everyday concerns with astounding ease and grace.

Perhaps because he possessed innate talent, Favs tended to separate people into two categories: those who had it and those who did not. I was lucky enough to be lumped into the haves. From the day I arrived he acted as if, all evidence to the contrary, his team benefited from having me around.

"So, is it amazing?" friends would ask.

Of course it was amazing. Sometimes Kathy, Valerie's assistant, would explain that we needed to reschedule a meeting because Valerie had been called into the Oval. She said this casually, as though her boss had been put on hold with the cable company and not summoned by the leader of the free world. Other times I would watch Favs and the POTUS speechwriters spitball lines for a set of remarks. A few days later, I would see those exact same lines on the front page of the *New York Times*. It was unbelievable. I felt like Cinderella at the ball.

Left unmentioned was that I also felt like Cinderella before the ball. I had never worked so hard in my life. In the speechwriting world, "holding the pen" means bearing responsibility for a set of remarks. My first week at the White House, I held the pen on seven speeches in five days. It was not uncommon for me to edit a single speech four times in an eight-hour span.

Also, the ratio between reward and risk was perilously skewed. If I did my job flawlessly, no one would notice. Fail to properly attribute a quote, however, or omit the *L* from *public investment,* and my mistake would be national news. Each day brought a once-in-a-lifetime experience, but each night brought intensely literal dreams about work. Tomorrow's speech left a senator unthanked. I had botched the phonetic spelling of *buenas noches.* I had failed to confirm a statistic with the relevant policy team.

The focal point—of both my workdays and my nightmares—was a black plastic rectangle the size of a deck of cards: my BlackBerry. By 2011, the BlackBerry was already becoming an anachronism. It was as if the entire federal government had signed up for Jazzercise, or gone to Blockbuster to rent *Space Jam* on VHS. It didn't matter. Like nearly every White House staffer, I lived and died by my device. Before long, I developed a Spidey-sense for incoming e-mail. My nerves would tingle, I would check the home screen, and then, a millisecond later, the red light on the top right corner would flash.

At the time, I didn't see this as a warning sign. It was a badge of honor. The moment a new message appeared, my thumbs leapt into action. I could read a chain of messages, reply to the sender, and relock the screen within seconds, all while carrying on a separate conversation face-to-face. I was an e-mail samurai.

I was also, however, a basket case. I would never have admitted it, but each new message triggered a fight-or-flight response. Every blinking light meant Valerie had another round of edits, or that a new set of talking points had to be composed ASAP. The only thing more frightening than a flurry of new e-mails was no e-mails at all. Just three minutes without feeling a vibration in my holster, and full-blown panic would set in.

Is there a malfunction? Did I miss a reply? Should I ping someone to ask why no one's pinging me?

I told myself I was handling the pressure. Then one afternoon, several hundred e-mails into my day, I yawned and something popped. Almost immediately, my right cheek ballooned. My face pulsed and throbbed. A few seconds of googling gave me a diagnosis: temporomandibular joint disorder. In layman's terms, I had clenched my jaw so tightly it had simply given way. Far more shocking than the injury, however, were my coworkers' reactions. When I described my condition to Jessica, a liaison in intergovernmental affairs, I assumed I'd get some sympathy. Maybe she'd even advise me to take the day off.

"Yeah, that happens a lot," she said, barely looking up from her computer. "First time for me was on the John Edwards campaign. Just don't chew for a while."

I followed her advice, and a week later my jaw no longer ached. But as my BlackBerry kept up its incessant buzz, I couldn't help but wonder if the White House had made a mistake. I had at least some talent. I was sure of that. Yet I also knew that there were more than three hundred *million* people in America. True, some of them were babies. But a lot of them were adults. It seemed unlikely that I was the best We, the People, could do.

ON THE LAST WEDNESDAY MORNING IN APRIL, I WALKED BY STRAUT'S office and noticed a crowd gathered around his TV. This was unusual. At the White House, televisions were almost always muted, the manic gestures of cable-news pundits a silent backdrop for actual work. Now, though, the volume was turned up full blast. In the center of the screen was President Obama, making an announcement from the briefing room. He looked profoundly annoyed. I turned to Alex, Straut's assistant.

"Holy crap! He's releasing his birth certificate?"

My introduction to the birther movement had come a few years earlier, through a blog written under the pen name "Texas Darlin." Texas was one of the few Hillary supporters who hadn't come around. Instead, it was as if Catholics and Protestants had split, and she'd responded by turning to Satan. Her posts made even Sarah Palin look restrained.

I kept tabs on Ms. Darlin's blog the way a coal miner keeps tabs on a canary. If she floated a conspiracy theory on Easter, by Christmas it would be part of the conservative mainstream. Still, when she began warning readers that Obama was born in Kenya, I was certain she had gone too far. *Who could possibly believe that?* I thought. And then, just a few months later, millions did.

If people like Texas Darlin kindled the birtherism epidemic, it was real estate mogul Donald Trump who fanned the flames. It seems unbelievable now, but in 2011, Trump's star was waning. His reality show was losing its audience. America was tiring of his brand.

Then, in a reboot of epic proportions, Trump began demanding the president's birth certificate. It was impossible to tell whether his act of showmanship was a triumph of strategy or instinct. Either way, it came with an adoring, built-in audience. A sizable chunk of red-state America was desperate for a hero, someone to unmask the foreignness of the president they loathed. Within weeks, Trump was leading a kind of anti-Obama revival movement. Instead of covering the administration's agenda, reporters began covering the birthers' antics. No wonder the president looked pissed.

And now he had called the Donald's bluff, posting his long-form birth certificate on the Internet for the entire world to see. Those of us standing around Straut's television unanimously agreed this was badass. In a single stroke, President Obama had just delivered a fatal blow to birtherism. He had exposed its most famous cheerleader as a fraud.

Best of all, he had done it just three days before he would see Donald Trump in person. That Saturday, they were both scheduled to attend the White House Correspondents' Dinner.

BECAUSE THE CORRESPONDENTS' DINNER PLAYED AN OUTSIZE ROLE in my White House career, I should acknowledge up front that it is bonkers. Imagine learning that once a year the British prime minister leads a mariachi band, or the Chinese premier performs burlesque. Yet in America, tradition dictates that each spring our commander in chief don black-tie attire and perform a comedy monologue in the ballroom of a Washington hotel.

In some ways, it's a fairly modern ritual. While every president since Calvin Coolidge has attended the dinner, most didn't tell jokes. Those who did kept their remarks focused on the relationship between the chief executive and the press.

Then, near the end of the Reagan era, reporters began inviting celebrities. Celebrities began showing up. In 1997, red-carpet buzz reached new heights when Ellen DeGeneres and her then-girlfriend Anne Heche arrived. In 2002, Ozzy and Sharon Osbourne accepted invitations at the peak of their reality-show fame. By the time President Obama took office, what was ostensibly a fund-raiser to provide scholarships for high school students had become a bilateral summit between

Hollywood and Washington, D.C. The night was nicknamed Nerd Prom, but Name Dropper's Paradise would have been just as accurate.

"As I overheard Chris Christie tell Sofía Vergara . . ."

"That reminds me of a joke I told the Secretary of Labor, Kim Kardashian, and two-thirds of the cast of *Glee*."

Each White House handles amateur night differently. Bill Clinton's staff fought tooth and nail to land their favorite lines, backstabbing colleagues without remorse. I once asked a Clintonite how to gracefully cut subpar jokes, and he told me not to bother. "If a joke really stinks," he advised, "push it into the middle of the highway. Let it get run over by a truck."

George W. Bush's team was less cutthroat, but also less confident in their principal's ability to perform. They often let goofy slides do the heavy lifting, while the president read captions that were difficult to screw up.

Obama's process was defined not by infighting or caution, but by an ironclad if unwritten rule. Everyone stayed in their lane. Jon Lovett was the president's funniest speechwriter—his color commentary was a fixture of our team meetings—so he was in charge of the jokes. Favs made edits and shaped the flow of the script. David Axelrod, already in Chicago for the reelection campaign, served as a kind of professor emeritus, offering advice, pitching one-liners, and initiating long e-mail chains full of puns.

Another dozen or so writers—some from the comedy world, others from politics—sent in jokes as behind-the-scenes volunteers. It was in this role that I had landed a few lines in the previous year's script. Now, however, I was on the inside. I was sure I had more to contribute. I hoped I would finally get my chance.

In many ways, I did. Favs and Lovett found me a ticket to the dinner. When we had to tweak setups or edit transitions, they welcomed my two cents. Also, because I could see each draft of the monologue, I knew in advance which lines would make the cut. As the day of the dinner neared, I had four jokes in the speech—not an overwhelming number, but respectable. I was especially proud of a line I'd written about one of the 2012 Republican candidates, in response to the birther mess.

"You may think Tim Pawlenty's all-American, but have you heard his full name? That's right: Tim 'bin Laden' Pawlenty."

I could picture the shock rippling across the audience, followed by a wave of laughter. It was going to be epic. I couldn't wait.

Still, the dinner wasn't the big break I was hoping for. Access to drafts aside, I found myself excluded from the joke-writing inner circle. I understood why Favs and Lovett didn't rush to include me in their Oval Office meetings. They didn't yet know if I could be trusted with information reporters would cut off a pinkie to obtain. But it stung to be so close and yet so far.

The day of the dinner, as I scoured Washington in a desperate search for cuff links, Favs, Lovett, and Axelrod met with the president in the Oval one last time. Not long after, the red light on my BlackBerry flashed. An updated draft. Some of the changes were hilarious, especially the long run of jokes that Lovett and Judd Apatow (one of our comedy-world silent partners) had written torching Trump. But as I scrolled through the latest version, I was stunned. They had butchered my best line. The reference to bin Laden had been deleted. In its place was the name of the Egyptian strong man deposed a few months before.

"You may think Tim Pawlenty's all-American, but have you heard his full name? That's right: Tim 'Hosni' Pawlenty."

Something inside me snapped. Yes, I was an entry-level speechwriter. Yes, everyone pays their dues. But this was absurd. How could they possibly think *Hosni* was funny? Were they too scared to have the president say the name of the world's most wanted terrorist? Did they not realize that was the entire point?

And besides, think of the phonetics. Say what you will about the man, but *bin Laden* has a ring to it. You've got the rhyming syllables on both sides, with a hard consonant in between them. *Hosni* is garbage. It starts out all flabby. It has no discernible end. Anyone could see that.

Even worse, when I complained to Favs, he told me the edit had come from the president himself. *Well, tell him to change it!* I thought. *Aren't you supposed to be the speechwriter?* For the first time since walking through the gates, I was 100 percent sure the White House needed my opinion. I whipped out my BlackBerry. Thumbs flying, I furiously typed a manifesto. I called up the menu of options. I placed my finger on the SEND button.

Then, abruptly, I paused. It was as if a tiny bureaucratic angel was whispering in my ear.

Stay in your lane. Staaaaay in your laaaaane.

Slowly, I holstered my BlackBerry. I went home and put on my tux. Then, ticket in hand, I took the bus to the Washington Hilton.

That night, I gawked at enough famous people to last a lifetime. I watched Amy Poehler search for her table. I witnessed Bradley Cooper mingle before taking a seat. I stared at Treasury Secretary Tim Geithner as he chose between finger foods at the reception. And from the back of the room, I watched President Obama's monologue, the best he had ever delivered. During the section on Trump, hundreds of Democrats and Republicans joined in bipartisan, mocking laughter. As the crowd applauded the president, the humiliated billionaire turned as red and angry as a blister.

Well, I remember thinking, *that's the end of this guy.*

The only downside of the night was Hosni. My joke fell flat, just as I had feared. I couldn't understand why no one else saw it coming. As the program ended, important people filed out of the ballroom to attend exclusive parties. I returned home to tell my friends about the time I used the urinal between Newt Gingrich and Jon Hamm. Still, for all the glamour, the night left me unsettled. I had been right about that punch line. I could have fixed it. Why didn't anyone listen?

The next morning, in a minor act of protest, I did something unthinkable. I turned my BlackBerry from vibrate to silent. Then I went to a music festival in Maryland with two friends from college, Nick and Claire. I wore a T-shirt and flip-flops instead of a suit. I drank a beer during daylight hours, even though Valerie had a speech tomorrow and I might be called upon to edit at any moment. For the first time since Straut told me I was hired, I felt twenty-four again.

All too soon, our day of freedom ended. I wasn't exactly thrilled about returning home. Glowering reluctantly in the back seat of the car, I switched my BlackBerry from silent to vibrate and it buzzed almost immediately. More talking points from Straut, I figured, or maybe a last-minute change to tomorrow's speech.

It was neither. To my surprise, the sender was Ben Rhodes, the president's chief foreign-policy writer. The subject line included the word *FINAL* and the acronym *UBL*. I was confused. POTUS wasn't scheduled to deliver a speech that evening. Also, I had no idea what those three letters stood for. Unresolved Banking Liability? Unanimous Bipartisan Legislation? Puzzled, I opened the e-mail.

Remarks of President Barack Obama
On the Death of Usama bin Laden
The White House
May 1, 2011

I was suddenly very glad I hadn't hit SEND the night before.

The news broke on Twitter a few minutes later. Nick, Claire, and I drove toward Washington, collectively freaking out. When we reached D.C. we tried watching CNN at a bar, but something felt off. Like moths responding to a stadium light, we each had the same instinct at once.

"Let's go to the White House!"

By the time we got downtown, Pennsylvania Avenue was packed with young Americans. College students waved flags and raised their arms in triumph. Bros hoisted fellow bros on their shoulders. Out of nowhere, people began singing the national anthem. Then they switched to an old chant, one I always associated with the Yankees winning a game.

Nah-nah-nah-nah, nah-nah-nah-nah, hey hey-ey, goood-bye.

I had never been part of such a raucous celebration, not even on Election Night. But looking from face to face, I didn't see joy or triumph. I saw relief. For people my age, 9/11 was the formative experience of our formative years. As kids, we were told America could do anything. Then a terrorist attacked our country and murdered thousands of our people, and we couldn't catch him no matter how hard we tried. For a decade, that failure cast a shadow on the promise we were raised to believe. Now, the shadow had been lifted. America had done what America set out to do.

Hey hey-ey, good-bye.

My BlackBerry vibrated. It was Valerie, with major changes to tomorrow's speech. She wanted to talk about the president's courage. She wanted to speak to his judgment, to his character, to the way those qualities had been present when she met him two decades before.

I hadn't ordered any raids. I hadn't been in the Situation Room, or even near it. But early the next morning, long after the crowds had left Pennsylvania Avenue, I would return. I would walk through the gates of the White House and do something, however small, that needed doing. In all likelihood, I was still not the best my country had to offer—my former TA could vouch for that. But in that moment, the question of whether I belonged seemed less important than the fact that I was here. In some small way, America was counting on me. In some small way, I was part of the team.

Still, though. Hosni? They really should have gone with Saddam.

4

THE CORRIDORS OF POWER

ALEX *(early twenties, staff assistant):* Can you help me with this e-mail?

DAVID *(early twenties, speechwriter):* Sure, what do you need?

ALEX: We're replying to a CEO. I don't want him to think we're blowing him off.

DAVID: Okay, but what *are* we doing?

ALEX: We're blowing him off.

In my five years at the White House, this was the first and only time my life resembled the television show *The West Wing*. That was not for lack of interest. Like nearly every Democrat under the age of thirty-five, I was raised, in part, by Aaron Sorkin. During my freshman year of college, my friends and I watched

West Wing DVDs on an endless loop, pausing only when our born-again roommate held pop-up Bible study in our suite.

> **MARK** *(deeply earnest):* Have you ever wanted to learn more about Jesus?
>
> **DAVID** *(equally earnest):* Have you ever seen the last episode of season two?

The witty banter. The brilliant grouches barging into the Oval Office. The satisfying conclusion each week. We couldn't get enough. But being introduced to politics by *The West Wing* was like being introduced to sex by *Debbie Does Dallas.* No matter how much more satisfying the real thing was than fantasy, in certain ways it was bound to come up short.

This was especially true when it came to office space. On the DVDs I devoured, the West Wing was spacious and grand. In real life the West Wing is a human ant farm that received its last major renovation in 1934. To be fair, the Oval Office looks like it does on television. A few of the president's top advisors occupy large, wood-paneled rooms best described as "robber-baron chic." But dozens of remaining staffers are squeezed into tiny outer offices, or piled clumsily on top of each other like dirty dishes in a sink.

The result was the world's most Type A sweatshop. At a time when Silicon Valley lured talent with ball pits and Ping-Pong tables, some of the most coveted jobs in Washington offered less personal space than Walmart during a Black Friday sale. Signs of overpopulation were everywhere. During peak hours, the line at the Navy Mess takeout window was so long you wondered if there might be a roller coaster at the end of it. In the ground floor that doubled as a waiting room, a faint odor of sewage occasionally infused the halls.

But the biggest lie of *The West Wing,* by far, was the walk-

and-talk. In President Bartlet's White House, staffers strolled side by side, trading barbs about policy while aides scurried in and out of the frame. In the actual West Wing, walking and talking was dangerous. One morning I left Favs's office and, perhaps distracted by the sewagey smell, nearly tripped over a pair of black leather shoes. I looked up to see who they belonged to: it was a face I recognized but couldn't quite place. I riffled through options. A new colleague? A B-list celebrity? A contestant on the latest season of *Top Chef*?

Oh, I realized, *I know where I've seen you before. You're the president of the United States!*

I leapt sideways just in time to avoid learning what the Secret Service does to people who head-butt Barack Obama in the chest. Still, the lesson had been learned. In the West Wing, possibilities were endless. But space was tight.

Which was one reason I didn't actually work there. Like most staffers, my office wasn't at 1600 Pennsylvania Avenue, but at 1650, in an ornate stone monolith known as the Eisenhower Executive Office Building. The EEOB sits inside the White House security perimeter, but in both size and style is nothing like the mansion next door. It's five stories tall. It occupies an entire city block. In 1888, Henry Adams, one of his era's most renowned curmudgeons, called the place "an architectural infant asylum." More than a century later, it's hard to argue his point. I regularly got lost inside my own workplace, adrift in a sea of imposing spiral staircases and stately marble floors.

If there were a *Hallways* magazine, the EEOB would surely score a centerfold. Six or eight of us could have strolled through the high-ceilinged corridors arm in arm like a cadre of Von Trapps. What we lacked was power. In fairness, our nearly six hundred rooms contained a handful of top presidential advisors. But the vast majority of us, myself very much included, weren't anywhere close.

Theoretically, no one noticed this divide. When we referred to the White House campus, we spoke of just one entity. "The building."

"Have you worked in the building since day one?"

"Is this ask coming from inside the building? If not, I'll ignore it."

"Why would you get lunch outside the building? It's Taco Wednesday!"

In reality, however, the White House campus held two distinct buildings, and everyone knew it. The pedestrian walkway between them, West Executive Avenue, took on almost mythic importance. My first day of work, I was handed a green plastic card and told to wear it around my neck. Straut pulled me aside.

"Don't worry," he said. His tone was low and secretive. "We're gonna swap that out for a blue badge."

This, I soon realized, was a big deal. A blue badge meant I could cross West Exec and Secret Service wouldn't stop me. Green-badge holders cross as well, but only with a blue-badged escort. This didn't quite make them second-class citizens. But there were certainly times when we blue-badgers treated them like Plain-Belly Sneetches from Dr. Seuss.

There was only one exception to our rigid caste system: a stray, tailless cat named Smokey. A malicious furball of indeterminate gender, Smokey patrolled West Executive Avenue with no dignity and no respect. Sometimes I'd hurry by a potted evergreen, late for a meeting, and Smokey would leap from hiding, hiss angrily, and disappear. Other times Smokey would ambush me near the checkpoint on the North Lawn. As I flashed my ID, what appeared to be a matted gray throw rug would bolt across my feet.

As far as I was concerned, Smokey was an intruder, so imagine my surprise when empty cans of Friskies and Purina began appearing along the wrought iron fence. At first I assumed these treats had been left by a kindly tourist. I later learned the truth: someone in the Secret Service was a cat person. Badge or no badge, Smokey was authorized to roam.

Perhaps this unique level of access explains why Smokey seemed to intuit our East-West office divide. When job numbers came in higher than expected, or we were winning the news cycle, I barely remember seeing Smokey at all. In those moments, blue-badgers flowed freely. If I had to bring a shirt to the dry cleaners, I'd take a shortcut through the Executive Mansion. If I was hungry, I'd pop over to steal a few of the M&M's that appeared, as if by magic, outside the Situation Room.

But when a crisis was brewing, or the economy showed signs of weakness, it was as though a drawbridge had been raised. Meetings in the Roosevelt Room or chief of staff's office were canceled. Officemates with business in the West Wing hunched between buildings in a sleep-deprived, self-important hurry.

It may be selective memory, but I'm almost certain that during these low points, Smokey guarded West Exec like a troll in a fairy tale. If I'm right, then I saw more of Smokey than ever during the debt ceiling crisis in the summer of 2011.

The debt ceiling is boring, complicated, and can wipe out your life savings overnight. For these reasons, it's worth taking a step back to consider it. Imagine buying something on your credit card. A blender, say, or a pair of those rubber shoe-glove hybrids that have mercifully gone out of style. At the end of the month, you pay your credit card company, you keep your ugly shoe-gloves, and your balance goes to zero. Simple.

But now imagine that instead, Congress gets to vote on whether or not you're allowed to pay your bill. You already own your regrettable shoes. You're not short on cash. But if a majority of lawmakers refuse to grant permission, it is nonetheless illegal for you to pay what you owe.

This arbitrary hurdle is the debt ceiling. The difference is that if you don't pay your bills, a pushy guy calls your house a lot. If America defaults on its debts, the global economy implodes.

The debt ceiling is an accident of history. There is no good reason for the world financial system to have a self-destruct button on Capitol Hill. But it does, and on May 10, 2011, House Speaker John Boehner threatened to press it. At a fund-raiser with wealthy Manhattanites, he acknowledged that triggering a default would be catastrophic. Yet he also warned that unless Obama cut trillions of dollars to federal programs, without raising taxes on the rich by even a penny, he would default nonetheless. His threat was almost comically absurd.

BANK ROBBER *(somehow straight-faced):* Violence is never the answer. Also, hand over the money or I'll shoot!

As late spring became early summer, Boehner's ransom demands occasionally came up in our West Wing meetings. But I never got the feeling we were truly alarmed. Because the debt ceiling was so dangerous, threatening not to raise it had always been a hyperbolic form of protest. It was the political equivalent of declaring that you'll kill yourself if Melissa from work posts another baby picture online.

Besides, it seemed only fair to cut the speaker some slack. A new, angry cohort of Tea Partiers had swelled the Republican ranks. By making extravagant threats, Boehner was throwing a bone to his most radical members. After letting his fellow Republicans shake their sillies out, he would surely join President Obama in a grand, bipartisan bargain to reduce the national debt.

If this was Boehner's strategy, however, it proved too clever by half. Only after he had begun his game of chicken did he realize that the right wing had cut the brakes. Instead of paring back their wish list, Republicans began adding to it. Eager to stay on the Tea Party's good side, Majority Leader Eric Cantor dropped out of negotiations he was supposed to lead. Throughout June and July, sillies were shaken. Wiggles were waggled. But nothing went away. Each day President Obama was presented with a choice that might as well have been written in letters cut from magazines.

IF YOU WANT TO SEE YOUR ECONOMY ALIVE, YOU WILL AGREE TO THE FOLLOWING DEMANDS.

It was around this time that Smokey became, in my mind at least, a constant fixture on the fringes of West Exec. Senior advisors stopped delivering remarks. Check-in sessions were scrapped. I was eager to write speeches, but Valerie was busy dealing with behind-the-scenes emergencies. She no longer had speeches to give. Each morning I'd display my fancy badge, walk down the grand corridor to my office, and have almost nothing to do.

As the crisis snowballed, I responded like any self-respecting millennial: I retreated into my phone. When I wasn't reading press clips about stalled negotiations, I passed the time with a game called Race Penguin. This sounds like a slur one Black Panther might hurl at another, but in fact the game's protagonist was a black-and-white bird, a cartoon one, who slid on his belly across different landscapes. If he got through the level fast enough, he lived to slide another day. If not, a polar bear ate him. As I tapped the screen, controlling my penguin's actions, I imagined his tortured existence, each moment a cocktail of monotony and stress. I could relate.

Race Penguin featured twenty-four levels, divided into three stages of eight. I completed the first stage, Ice World, around the time Republicans added an unworkable balanced budget amendment to their ransom demands. As I polished off the second stage, Desert World, Democratic leader Harry Reid made a major concession. Tax increases would no longer be required as part of a final deal.

By July 25, I was on the game's very last level, the final section of Rainbow World, and President Obama had no choice but to deliver a prime-time debt-ceiling address. It was a tense moment for the president. The economy was less than two weeks from collapse. It was also a tense moment for me. My penguin was stuck. No matter how flawless his performance, how well-timed his slides, the polar bear always caught him. I could reach only one conclusion. The game was broken. It was impossible to win.

"TONIGHT, I WANT TO TALK ABOUT THE DEBATE WE'VE BEEN HAVING in Washington over the national debt."

The president's address to the nation was the kind of speech reporters were already calling "Vintage Obama." Sober. Smart. Postpartisan. "The American people may have voted for divided government, but they didn't vote for a dysfunctional government," he said.

Watching the livestream from my office, headphones in my ears and feet on my desk, I was sure the remarks would move the needle. Two and a half years earlier, one of Obama's speeches had changed my life. Surely tonight's speech could change public opinion and bring Congress to its senses.

A few days and several hundred penguins later, President Obama announced he would cut more than two trillion dollars from the budget, while getting essentially nothing in return.

The moment the news broke, Jon Carson, the Office of Public Engagement's executive director, called a meeting in one of the EEOB's handsomely decorated conference rooms. Ensconced in leather chairs, surrounded by portraits of severe-looking dead white men, we went over our talking points. Most of the cuts weren't immediate. Most social safety-net programs had been spared.

But was I wrong to think the stern, pasty faces on the walls looked skeptical? As hard as we tried to frame the debate, the most important point was the one that went unsaid: if you have to hold an hour-long meeting about why you won, you didn't.

The debt-ceiling deal did nothing to pacify Republicans in Congress. If anything, the situation grew worse. "It's a hostage that's worth ransoming," declared Mitch McConnell, the GOP leader in the Senate, promising more brinkmanship to come.

I longed to wipe the self-satisfied grin from McConnell's turtlish face. If I was being honest, however, my anger at the Tea Party was nothing compared to my disappointment in the president who caved to its demands. From the moment I laid eyes on my candidate, what drew me to him was not just that he was on the right side of history. It was that he knew how to win. Sure, there would be small setbacks, but when the stakes were highest, Obama would prevail. Not this time. In his biggest fight yet with the Republican Congress, with trillions on the line, he had come up short.

People in Washington talk about disillusionment the way people in high school talk about virginity. Your most mature peers have already gone all the way, you're told. If you haven't done it yet, it's only a matter of time. Maybe you'll settle down as a career public servant, remaining as pure as possible. Maybe you'll join a lobbying firm and sleep with anything that moves. It doesn't matter. The point is that there will be a single, defining moment, and it will close the door on your childhood forever.

Except that's not how disillusionment works. In the wake of a lopsided debt deal, I didn't suddenly decide President Obama was a fraud. I had no impulse to yell about betrayal while sweeping papers off my desk. Instead, I was struck by even more worrisome questions. What if we're all just race penguins? What if the game is broken? What if we do everything right, and the polar bears still eat us in the end?

DURING THE WEEKS THAT FOLLOWED, AS NEWS REMAINED GRIM and our approval ratings struggled to recover, friends and family grew concerned.

"What's the feeling like over there?" they wondered. They asked this as if there were some special emotion that exists nowhere but the White House, frustropefulness, maybe, or chaostalgia.

Other times they would frame the question more personally. "Is my buddy Joe okay? How's Barack holding up?"

To my surprise, my friends and family were disappointed when I told them I had no idea. Now that they knew a White House staffer, they assumed they had direct access to the president of the United States. Sitting in my EEOB office one afternoon, I received a text from my little sister. *How come the department of Homeland Security doesn't have a mailing address?* Even under the best of circumstances, this is a disconcerting question to receive from a family member. But if you work in the White House you want to have an answer. I didn't.

It was like that with everything. Grandpa Irving e-mailed to ask if "my people" might take a look at his plan for a national network of water pipelines. Uncle Gabe warned me that Obamacare was causing black-market health care clinics to pop up left and right. Most of all, everyone had one question.

"So, have you met Obama yet?"

"No, not yet," I would say, hoping to change the subject. Instead, I'd get a look, which I soon learned means, "You may be twenty-four years old and working at the White House, but you're still a disappointment to your family and friends."

I couldn't blame them. They assumed, just as I had, that the White House is like *The West Wing,* where everyone hangs out with the president. Either that or it's like *Scandal,* where everyone has sex with the president. But really, the White House is like the Death Star. There are thousands of busy people running around, each trying to make their own little piece of the ship function the way it's supposed to.

"Just because Darth Vader is the public face of the organization doesn't mean every stormtrooper gets one-on-one time," I'd explain. This never worked.

Frankly, no one was more disappointed than I was. No one wanted me to meet Obama more than me. There were two reasons for this. The first was corny, but true: I believed there must be something I could do for my country, even if I didn't yet know exactly what it was. I thought I could be the kind of person who makes the president a tiny bit better at his job, just by being in the room.

The second reason is that I wanted Barack Obama and me to become best friends. I'm not saying I assumed this would happen. None of us did. But anyone who worked in the Obama White House and tells you they didn't imagine becoming buddies with the president is either lying or named Michelle. Every so often you'd hear stories. Someone from the Office of Management and Budget got a fist bump in the hallway. A staff secretary was invited to play cards on Air Force One. The moral was obvious. Any moment could change your life forever.

My first chance at a life-changing moment came in November 2011, when Favs asked me to write the Thanksgiving video address. If the State of the Union is on one end of the speechwriting spectrum, "Happy Thanksgiving, America!" is firmly on the other. Still, I was floored. I had written jokes for the president, but this was different. Here was an opportunity to write something profound, something unique, something all-

American. I researched past Thanksgiving videos. I read essays about Pilgrims. I went through draft after draft.

Finally, on the day of the taping, I made my way to the Diplomatic Room. It was one of the most beautiful rooms in the residence, the walls covered in a wraparound mural of nineteenth-century American life. By the fireplace, a wooden chair had been set up for the president's arrival. I stood as far from it as I could. Behind the camera, a woman in a vest and button-down shirt was adjusting the focus.

"First time here?" she asked. I tried to sound nonchalant and casual. Instead, I immediately cracked.

"Yes! Help! What do I do?"

This was how I learned that Hope Hall, the president's videographer, was a deeply calming presence. A rare free spirit in a building full of joiners, she smiled serenely. Then she told me not to worry. All I had to do was wait.

So I waited. And waited, and waited, and waited. Finally, just as I was beginning to wonder if the whole thing was a nightmare or a practical joke, one of the A/V guys got an e-mail on his BlackBerry.

"He's moving."

There was a kind of crackling in the air. Then, a minute later, President Obama walked through the door. Suddenly, my nose itched. Was I allowed to sneeze? Had I turned off my phone? Did I have loose change in my pocket? I was struck with an overwhelming desire to shift my weight from leg to leg.

The president was standing up, so we stood up. He sat down, so we sat down. He looked at the camera, but before he could begin taping, Hope stopped him. "Mr. President, this is David," she said. "This is the first video he's ever written for you."

President Obama looked at me.

"Hey, David," he said. "How's it going?"

I had exactly one thought in that moment. *I did not realize we were going to have to answer questions.* And I have no idea what happened next. I literally blacked out. I went home for Thanksgiving, and my family said, "Have you met Obama yet?" and I said, "Yeah," and they said, "What did he say?" and I said, "How's it going?" and they said, "What did you say?" and I said, "I don't know, I blacked out."

Silence. That disappointed look.

I understood their disappointment. If I was going to make the president better at his job just by being in the room, I would have to answer far more difficult questions than "How's it going?" There was no indication I could do it.

At least my family could take pride in the video I had written. My grandmother on my mom's side was particularly thrilled. She was with her boyfriend, Bill, ten years her senior, and together they made an excellent team. Thanks to the accumulated effects of time and Chivas Regal, Grandma would forget she was repeating herself. This was fine with Bill, who hadn't heard her the first time. Their shared love of shouting, combined with short-term memory loss, made them a kind of two-person hype squad.

"Can you believe it? David wrote something for the president."

"What?!"

"DAVID wrote a VIDEO! For the PRESIDENT!"

Even this accomplishment was short-lived, however. For all the effort I put into my script, something had been missing, and Fox News found it. It was remarkable how quickly they worked. The first headline was up before the turkey was even carved: OBAMA LEAVES GOD OUT OF THANKSGIVING ADDRESS. "Nowhere in the 11-paragraph address does he mention the Almighty," the article said.

By any fair measure, this was nonsense. The president had used the word "blessing" in his video. Who did Fox News think was handing out those blessings? Oprah? But the damage had been done. Right-wing media had ginned up a controversy, and respectable news outlets now felt free to cover it. OBAMA LEAVES GOD OUT OF THANKSGIVING SPEECH, RILES CRITICS, read the headline on ABC News.

I returned to work Monday expecting to be widely shunned. To my surprise, nearly all of my speechwriting colleagues had a Fox News story of their own. "Don't be too hard on yourself," they said. But my coworkers' graciousness couldn't mask the fact that I had failed. I wanted to do something for my country. I wanted to be friends with Obama. I was no closer to either goal.

If I ever got a second chance, I promised myself, I wasn't going to squander it.

To my surprise, the opportunity came just a few weeks later. I was sitting in my office when Favs called. "Betty White is turning ninety years old," he explained. "NBC is doing this special where famous people wish her a happy birthday, and you're pretty funny, and no one else wants to do it. Want to give it a shot?"

Of course I did. This was my Gettysburg Address.

The taping was on Friday, and we had a week to make things perfect. Jon and I started by coming up with a joke. As the president signed a birthday card, the audience would hear his message as a voice-over:

> *Dear Betty, you're so young and full of life, I can't believe you're turning ninety. In fact, I don't believe it. Please send a copy of your long-form birth certificate to 1600 Pennsylvania Avenue, Washington, D.C.*

Step two was to purchase the card itself. There was a CVS a half block from the White House. I studied their Hallmark display as carefully as a homicide detective at a crime scene. Then, just when I was about to make my purchase, I had a stroke of genius. We were going to film the president twice, from two different camera angles. But during the second shot, we couldn't let anyone see that he had already written his birthday note. In other words, we didn't need just one card. We needed two!

Yes! I thought. *This is how White House staffers are supposed to feel.*

I proudly returned to my office, certain I had saved the day, and began working on the final joke for the skit. Finally, I came up with something. As the video ended, President Obama would put in headphones. Then he'd pretend to listen to the theme song from *The Golden Girls,* Betty White's most popular show.

As the week drew on, I practiced explaining the joke to the president. I found the perfect headphones, a pair of white earbuds that would look great on camera. I listened to the *Golden Girls* theme song on repeat, just to get in the mood. By the time Friday dawned, I was brimming with confidence.

Then came the phone call. "All right, head on over to the West Wing."

Perhaps there are some people who, summoned to the Oval Office for the very first time, walk in there like it's no big deal. Those people are sociopaths. For the rest of us, attending your first Oval Office meeting is like performing your own bris.

To make matters worse, when you have a meeting in the Oval Office, you don't just go into the Oval Office. First you wait in a tiny, windowless chamber. It's kind of like the waiting room in a doctor's office, but instead of last year's *Marie Claire* magazines they have priceless pieces of American art. And instead of a receptionist, there's a man with a gun. And in a worst-case scenario, the man with a gun is legally required to kill you.

It turns out this little room is the perfect place to second-guess every life choice you have ever made. As Hope Hall, the videographer, joined me on the small couch, I silently approached a nervous breakdown. Did I remember how to explain the concept? Did I have both cards? Were the headphones still in my pocket? How about now? How about now? I was on the verge of losing it completely when one of the president's personal aides emerged.

"Okay. He's ready for you."

To my credit, the first time I walked into the Oval Office, I did not black out. In front of me I could see a painting of the Statue of Liberty by Norman Rockwell. Behind me, out of the corner of my eye, I could see the Emancipation Proclamation. Not a photocopy or poster. The. Emancipation. Proclamation. I didn't turn to look at the document, but I could feel the message it was sending through the room.

"I'm here because I freed the slaves," it seemed to say. "What are *you* doing here?"

Behind the giant, wooden *Resolute* desk sat President Obama. Judging from his expression, he, too, might be wondering what I was doing here. But I wasn't worried. I had spent an entire week practicing how to explain this video. I stepped forward. I opened my mouth to speak. And face-to-face with the leader of the free world, what came out sounded like an exchange student about to fail an exam.

"Betty White?!" I heard myself say. "Card and . . . birthday? Sing song theme *Golden Girls* headphones video, yes please?"

President Obama looked confused. Hope jumped in, rescued me, and began filming, but I nonetheless felt concerned. This was my chance to show the president that I was a consummate professional. In my professional opinion, we weren't off to a great start.

Still, I knew I had another opportunity: my second birthday card. I was about to show the leader of the free world that I had saved the day. As soon as Hope captured her first shot, I strode to the desk, surprising even myself with my confidence.

"Mr. President, we're about to film a second shot from a different angle, but we want it to look like you're writing your birthday greeting for the first time," I explained, reaching into my jacket. "I'll need to take that birthday card and replace it with this identical one."

President Obama cocked his head. "We're shooting from all the way on the other side of the room?"

"Yes, that's right."

"So no one can actually see *inside* the card."

"Yes. That's right."

"So, I can just pretend to write in the card I have now. I don't need a second one at all?"

"Yes . . . that's right."

I returned the backup card to my pocket. Strike two.

Still, I wasn't giving up. I had one final shot, the joke with the earbuds, and I was determined to make it count. The moment Hope was done filming, I walked back up to the desk, reached into my pants pocket, and withdrew what looked like a hair ball made of wires.

I still don't know what happened. Somewhere, probably in that waiting room, I had worried this thing into a hopeless tangle. Now I had no idea what to do, and so I did the only thing I could think of: I handed the whole pile to the president of the United States.

If you work in the White House, you will hear a saying: *There is no more valuable commodity on earth than a president's time.* I always thought it was a cliché. Then I watched Barack Obama untangle headphones for thirty seconds, while looking directly at me. He untangled and untangled. Finally, the president turned to Hope and sighed.

"Shoddy advance work."

He said it in a way that let me know that (a) he was just joking, and (b) he was not even a tiny bit joking. And in that moment, my heart sank. This was my third chance to make a second first impression on the president. I had let myself down. President Obama asked a question, but I heard it only faintly, as though a layer of gauze had been placed between us.

"I should probably bob my head back and forth as I'm listening. Wouldn't that be funnier?"

"Yeah, it would," I replied. But there was no rescuing my life-changing moment. I was in the Oval Office with the president, and all I wanted was to slink away. I stood in silence while Hope readied the final scene, knowing I would never get another chance. President Obama looked toward the camera.

And then, he paused.

"Hang on," he said. "If I'm going to bob my head in time to the music, I need to know how the music goes. Does anyone know the *Golden Girls* theme song?"

There was silence. President Obama looked at Hope. Hope didn't say anything. I looked at Hope. Hope didn't say anything. So President Obama looked at me.

And suddenly, I knew exactly what I could do for my country.

I planted my feet on the Oval Office carpet. I cast a brief glance at the Emancipation Proclamation behind me. Then I looked the commander in chief straight in the eye, and I began to sing.

Bah-bum-bum-bum, thank you for being a friend. Bah-bum-bum-bum, traveled down the road and back again. Something, something, you're a pal and a confidant.

Patriotically, enthusiastically, I continued.

And if you threw a party, invited everyone you knew-ooh-ooh . . .

President Obama gave me a look that indicated, politely but firmly, that we were encroaching on the president's time.

But it worked! The president bobbed his head in time to the music. NBC got their video. Betty White got her card. I left the Oval Office that day with my head held high, knowing that the president was just a tiny bit better at his job because I was in the room.

It's not as if I shed all my doubts that afternoon. I still wondered if we were race penguins. I still worried our hard work might never pay off. But as 2011 drew to a close, I felt a renewed sense of confidence. Our approval ratings were heading in the right direction. Smokey was nowhere to be seen. And I finally had an answer to the question on everyone's mind.

"So, have you met Obama yet?"

"Met him? Well, I don't want to brag or anything. Let's just say I'm thankful he's a friend."

5

THE SALMON IN THE TOILET

I never set out to become a connoisseur of White House men's rooms. It just kind of happened. You move into a new apartment, get to know the area, and one day, to your surprise, you have strong feelings about every pizza place in the neighborhood. That's what I went through, only with porcelain and liquid soap.

My favorite restroom was on the ground floor of the residence, next to the ceremonial library. The white marble floors and sinks radiated luxury, yet their muted, opaque quality kept them from showing off. "It's honestly no big deal," they seemed to say. "Some of us were just born pretty." The overall effect was both stunningly impressive and refreshingly humble, the best of American democracy superimposed upon a WC.

West Wing men's rooms offered unique charms of their own. For proximity to power there was the stall tucked against the Roosevelt Room, just footsteps from the Oval. For retro quirkiness, there was the restroom across from Valerie's office. Urinals there were large and basinlike, like bathtubs sawed in half. (Even stranger, they flushed via bulky foot pedals placed twelve inches from the floor.) At ground level, the bathroom near Favs's office boasted the building's only shoe-polishing machine, the kind that looks like two Muppet scalps attached to opposite ends of a stick.

Not surprisingly, the facilities in the EEOB were less distinguished. If anything, their defining feature was a trigger-happy automatic flush. I won't go into too much detail. All I'll say is that I was the frequent victim of an impromptu bidet.

Yet despite these shortcomings, when it came to personal significance, no restroom could match the one in the southwest corner of the EEOB's ground floor. This was because of something unforgettable (in advance, don't worry, not gross) that happened six months after I started my new job.

It was a special time. I could finally navigate my surroundings, but the novelty of the building had yet to fade. Even the most routine pee break glowed with history. Descending a spiral staircase to the ground floor, I remembered that FDR had worked here during the 1910s. That was before polio claimed his mobility, and I imagined his shoes clicking and clacking on the steps. The EEOB was also home to the vice president's ceremonial office. Tugging on the engraved metal knob of the men's room door, I wondered if Nixon or Johnson had ever strained against its weight.

Then I entered the bathroom, and found the sole urinal occupied by someone in a bulletproof vest. This too was a kind of wonder. For the vast majority of my twenty-four years, I would have been stunned to see a real-life Secret Service agent. Diving in front of bullets. Driving through red lights. Sniping bad guys from rooftops. They were as mythical as X-Men, and no less devoted to saving the day.

Now, after a few months at the White House, I still thought of Secret Service agents as heroes. But they were also people I peed next to. Not wanting to wait for the agent at the urinal, I scooted past him and opened the door to the stall. Stepping inside, I closed the latch behind me and turned around.

And that's when I saw it: a fillet of grilled salmon, unblemished by a single bite mark, sitting in the toilet bowl.

This was not the most historic thing I witnessed at the White House. It was not the most profound. But it was, without question, the most remarkable. Think about it. How many people have met Barack Obama? Tens of thousands, maybe even hundreds of thousands. How many people have found a salmon in the toilet at work?

I have. And because I have, I can assure you the experience raises more questions than it answers. For example, why aren't there any side dishes? What would happen if I flush?

But those questions would come up anywhere. As I began the next phase of my inquiry, the search for suspects, a simple piece of fish took on far greater meaning. I knew it was an inside job—Ike's was serving salmon that day—but beyond that I was stumped. The National Security Council offices were only three floors above me. Was someone there really so woefully incompetent? Or maybe it was the econ team. Had a person responsible for billions in federal grant money simply snapped? And let's not forget the man in the bulletproof vest I had spotted just feet from the crime scene. Could a Secret Service agent have gone rogue?

This, I was learning, is the power of the White House: it sprinkles its significance onto anything nearby. A staircase becomes more than a staircase. A doorknob becomes more than a doorknob. A toilet-salmon becomes more than a toilet-salmon. It's astonishing to behold.

TO APPRECIATE THE FULL EFFECT OF THIS WHITE HOUSE FAIRY DUST, all I had to do was invite friends to bowl. *The Truman Alley* sounds historic, elegant, and fancy. In reality, it's a dump. With its stain-resistant carpeting and oversize industrial sink, the cramped outer room appears designed for autopsies. The only decorative touches—framed photos of presidents bowling—are easily available on Google Images. Also, there are only two lanes. For the vast majority of my White House tenure, one of them didn't work.

Worst of all, the only way to reach Truman Alley is through the EEOB basement, a warren of exposed wires and flickering fluorescent lights I am fairly certain was the setting for a *Saw* movie. After every bowling excursion, I led visitors through the murdery labyrinth to the surface, then braced myself for their disappointment. My guests struggled to voice their emotions. The only thing worse than their silence was the outrage I was certain would follow. Finally, after composing themselves, they would look me in the eye.

"That was amazing! Thank you so, so much!"

Where I saw the world's shabbiest rec center, they saw an exclusive pleasure paradise. That's what the White House can do.

It was only a matter of time before a question, shameful but unavoidable, crossed my mind. How much of that fairy dust rubs off on me? And more specifically, what could it do for my dating life?

For those who already had game, a White House job made seduction almost painfully easy. There was the winter, for example, when a blond local newscaster caught the eye of a coworker. (I'll call him Chase, because that's what he enjoyed.) Out of nowhere the anchor began receiving invitations: to a holiday reception in the residence; to a sports team's visit in the East Room. Each time she arrived, who should be seated beside her but Chase? He'd charm her for a few minutes, drop a couple of names, and then apologize for being so busy he couldn't stay.

It was almost too easy. After sealing the deal, Chase bragged about his conquest, but anyone could tell he was just going through the motions. He sounded like a big-game hunter given permission to shoot elephants at the zoo.

On exactly one occasion, at a summertime cookout in a friend's backyard, the White House worked a similar magic on me. I was in line for a beer refill when I bumped into Rachel, an activist type I knew from school. As usual, she didn't pay me much attention. Then I mentioned my new job. Suddenly, she was transfixed.

"Do you have a business card? Can I see it?"

I reached for my wallet. As she ran her finger over the presidential seal, I saw her eyes mist over. Then they suddenly narrowed, as the image in her field of vision made contact with a fantasy in her head.

"It's no big deal," I said. "It's only a card."

Unfortunately, the magic didn't last. After just two dates, and an equal number of lackluster make-out sessions, Rachel called it quits. I didn't need to ask what happened. Blinded by my business card, she had leaned in to kiss Rob Lowe's character from *The West Wing* and instead made contact with me. Even I couldn't help but feel sorry for her.

And yet, as unsatisfying as this encounter was, it was the only time my job title improved my love life. Working at the White House made me no more seductive. If anything, meeting people was harder than ever, since I was constantly at work. A few months after receiving my first box of business cards, I threw in the towel and took the online dating plunge.

I once heard Chase refer to OkCupid as "practice." But I

approached composing a profile with the same intensity I devoted to writing Valerie's remarks. What was the subtext of my favorite movies and TV shows? Were my "Six Things I Can't Live Without" in the perfect order? Most important, how could I make myself appealing while disclosing as little personal information as possible? I know that sounds paranoid. It *was* paranoid. It was also smart. With just a year until the election, even the lowliest presidential staffers were targets for the conservative press.

It hadn't always been this way. For most of American history your private life stayed private, even if you were commander in chief. Even after Bill Clinton's escapades made headlines, it was understood that the young were off-limits until they reached trophy size.

But by late 2011, inspired by the success of Fox News, new right-wing outlets were exploding like spores from a fungus. It wasn't just *Breitbart* and *Glenn Beck TV*. The *Washington Free Beacon* combed through Facebook pages of young Democrats, searching for embarrassing pics. Something called Project Veritas acted as a far-right candid camera, baiting progressives into saying something stupid, then releasing the heavily edited results. It was open season on everybody.

Determined to stay out of the crosshairs, I left my dating profile aggressively vague. This was particularly difficult on OkCupid, which relies on hundreds of questions to surface a match.

What's your relationship with marijuana?

Pass.

Which is worse, book burning or flag burning?

No. Absolutely not.

How willing are you to try new things sexually?

Far more willing than I am to answer this question, that's for sure.

I was a blank slate, and it was hardly surprising when my online matches and I failed to hit it off in person. Then, to my surprise, I came across a profile nearly as vague as mine. It belonged to someone I'll call Nora. When I sent her a message, she explained that she, too, had joined OkCupid to avoid meeting people at work. We made a date for the following week, at an underground bar on Connecticut Avenue. As usual in Washington, we started by comparing careers.

NORA: I'm in communications.
ME: That's so funny, I'm also in communications! In the private sector or government?
NORA: In the administration. How about you?
ME: Small world! I'm in the administration, too! Which agency?
NORA: Well, actually, I work at the White House.
ME: Wait, what?

Not only did we belong to the same department. We worked in the same building. On the same floor. Nora's office was down the hall.

What would have been cute in a romantic comedy felt icky and overly personal in real life. That was our first and only date, but from then on we'd administer clipped, uncomfortable nods in the hallway, as though we'd run into each other at a porno theater the night before. I was through with online dating. I was ready to give it up.

And then I met Jacqui. I found her attractive, of course. But even more than her looks, I was struck by her sense of purpose. She seemed fiercely determined, even just crossing the street. Halfway through her third year of law school, she nonetheless managed to make it through an entire evening without once using the phrase *according to the law*. She told me about growing up in New Jersey, and yet, despite my New York City background, I found her impossible to hate.

About an hour into our first date, a waiter came to take the drinks menu. Jacqui smiled and shot out a hand. "We're keeping that," she said. Her tone was warm and charming, while also suggesting someone might lose a finger if he didn't obey. My favorite TV show is *The Sopranos*. I've always had an odd thing for Laura Linney. I was floored.

IN MY PERSONAL LIFE, I WAS BEGINNING MY FIRST SERIOUS RELAtionship in years. At work, I was going through what can only be described as a slutty phase. I was, of course, Valerie Jarrett's speechwriter. But Bill Daley, then the White House chief of staff, had also been told I was *his* speechwriter. Straut and his fellow senior staffers were told I was the senior staff speechwriter. Three years out of college, I wrote six commencement addresses for five different White House VIPs.

In the *House of Cards* version of my life, my exhausting schedule would have been rewarded with influence. Slowly but surely, I would turn speakers against each other, putting words in their mouths in the service of some nefarious end. But that's not how speeches work. In the real world, speechwriters are more like personal trainers than puppet masters. They can help you present the most attractive version of yourself to the public. They can't turn you into someone you're not.

Nor can they "find your voice." This is the most common misconception about speechwriting. It came up especially often in the years before I started at the White House, when I wrote for CEOs. "You seem capable," they would tell me, "but can you really find my voice?"

"I think I can manage it," I'd reply gravely.

Left unsaid is that it would be easy, because when it comes to rhetorical styling, 99.9 percent of speeches sound the same.

Martin Luther King had a voice. John F. Kennedy had a voice. With all due respect, you probably don't.

What you do have are thoughts. What you need, although you may not know it, is someone to organize them. A good writer can take your ten ideas and turn them into one coherent whole. Where you see two all-beef patties, special sauce, lettuce, cheese, pickles, and onions on a sesame seed bun, a speechwriter sees a Big Mac.

Just as important, speechwriters act as stand-ins for the audience. If you're working with a writer, chances are you're an expert. And if you're an expert, chances are you're boring. You can't help it. The more you know about a subject, the harder it is to express your knowledge in a way the rest of us understand. As professional dilettantes, speechwriters use their short attention spans to your advantage. Sifting through the dense muck of fact and anecdote, they find nuggets worth something to the outside world.

This doesn't always turn out perfectly. Sometimes the final draft will forgo a single thesis in favor of a seven-point list, causing eyes to glaze over. (If the list is divided into sublists, as occasionally happens, the remarks can legally be used during surgery as anesthesia.) Other times the speaker insists on heaps of esoteric verbiage, then blames the writer when audience members start checking their phones mid-speech. But if you're holding the pen, and everything works as it's supposed to—if you define the forest of an argument without losing sight of the trees—the speaker will be forever grateful for your work.

"Wow, you really found my voice!" they'll say.

And you'll say thank you. If you were the kind of person who enjoyed correcting powerful people for no reason, you would have gone into a different line of work.

Because that's the other thing about speechwriting: unlike novelists or poets, the speechwriter must let go. Take prepositions.

Valerie followed the rule that you could never use one to end a sentence. I found this unnecessary. In my own writing, preposition placement isn't something I lose sleep over. But this was not my own writing. It belonged to Valerie, and the prepositions rule was one about which I did not argue.

The same went for "At the end of the day." I've always liked the singsong nature of the phrase, but each time I added it to a draft, Valerie cut it. That this was a matter of preference rather than grammar was unimportant. Speechwriters and speakers are free to disagree on just about anything. But at the end of the day, the speaker is always right.

Besides, Valerie's remarks were far easier than Bill Daley's. With the chief of staff, the challenge was not word choice but pronunciation. A compact, broad-shouldered man, Bill had developed the habit of speaking with his chin tucked into his neck. With most words this wasn't a problem. *S* sounds, however, lodged themselves in his Adam's apple before tumbling unceremoniously from his mouth. I tried to skirt this obstacle by avoiding sibilants at all costs. But consider the sorts of sentences Bill might be expected to say.

"As the president's chief of staff, I assure you he takes the possibility of rising deficits seriously."

"One of the president's most successful and courageous actions was his decision to send in Seal Team Six."

Sometimes, speechwriters must surrender to circumstance.

There was one final element to my professional promiscuity: I was beginning to write more often for the president. In the summer of 2011, there were eight of us who wrote POTUS remarks. Like many speechwriting teams, we were all below the age of forty. (At the time, we were also exclusively white and male, a regrettable trend my hiring did nothing to reverse.)

There was no "education writer" or "jobs writer." Still, each member of the team had a niche. Favs held the pen on major economic and political addresses. His deputy, Adam Frankel, handled civil rights. Cody Keenan (who became deputy once Adam left) took tragedies and the middle class; if it belonged in a Springsteen album, Cody was on it. Along with jokes, Jon Lovett was responsible for all things science and technology. Ben Rhodes and Terry Szuplat handled foreign policy. Kyle O'Connor, the youngest POTUS speechwriter, took whatever everyone else didn't want.

I took whatever Kyle didn't want. This meant I was not exactly writing for the history books. A brief ode to Puerto Rico. A shout-out to some longtime Chicago friend. A few kind words for the Hispanic Chamber of Commerce. But if my remarks weren't destined for greatness, they did help me learn President Obama's verbal tics. When delivering praise, for example, he avoided *good* and *great,* preferring synonyms that were ever-so-slightly formal.

> "Our *fantastic* attorney general, Eric Holder, is in the house."
>
> "I want to thank Congresswoman Chu for her *outstanding* work on this issue."
>
> "Senator Klobuchar is doing a *tremendous* job."

He had other preferences as well. Some speakers subscribe to an old bit of wisdom: "Tell 'em what you're gonna tell 'em, then tell 'em, then tell 'em what you told 'em." But President Obama was more lawyerly than folksy. He preferred to make one argument start to finish, to tell a single story with a beginning, middle, and end.

Then there were the ad-libs. Unlike Vice President Biden, who regularly gave his teleprompter operator a heart attack, POTUS didn't often go off script. Occasionally, however, he'd cock his head slightly to the side and pause. It happened so quickly I think only the speechwriters noticed. But if you knew what to look for, you could see a dialogue play out in his head.

> You know, this line isn't cutting it. I bet I could come up with something better.
> ***Are you sure? Maybe the speechwriter knows something you don't?***
> He's how old? Twenty-five? Seems unlikely.

Then the president would deliver something on the fly and, nine times out of ten, the crowd would break into applause. More than any other speaker I've seen, President Obama thrived on this enthusiasm. On the rare occasions he didn't receive it, it reminded me of one of my bad OkCupid dates: POTUS and the crowd, going through the motions, trapped in a downward spiral of disinterest. When the speech finally ended, he would give a short, stoic nod, the kind that says, "Well, moving on."

But other times, right after his remarks were finished, President Obama would give the podium a satisfied little thump with his right hand. This was his way of declaring victory, a quiet but unmistakable "nailed it." These were the speeches where the crowds started off cheering, and the cheering only grew as the remarks went on. When the audience was with him, POTUS found a gear no other speaker could match.

That's why, if my OkCupid profile had included the question, "What's the most important thing about remarks for President Obama?" I would have said this: Write long sentences. Most speakers can't handle them. They need to keep things tight. Otherwise they get lost. But Barack Obama could control a run-on sentence the way a sports car makes turns at speed, emphasizing, pausing, finding beats within the words and phrases not because of the punctuation but thanks to his innate talent as an orator, his voice rising and falling and carrying you along with it, so that by the time he reached his final crescendo you felt bigger than yourself, and better than yourself, and part of and proud of and lucky to be alive in the greatest country on earth.

POTUS speeches were fun.

They were also, in theory at least, the best way for the president to influence the public he served. One of the most important moments in the 2008 campaign had come during the primaries, after ABC News uncovered video of Obama's pastor making racially charged remarks. The candidate's response to the scandal? A sober, thoughtful, thirty-eight-minute address. The remarks were titled "A More Perfect Union," but were soon known as "The Race Speech." Their resounding success set the bar for every bit of rhetoric that followed. In White House meetings, it even became a running joke.

"We need a race speech for Greek Independence Day."

"We need a race speech for congratulating the winner of the Stanley Cup."

But beneath the sarcasm lay a growing sense of insecurity. More than ever, it seemed that even the most well-crafted presidential address was powerless to change America's course.

GRIDLOCK. THAT WAS THE SHORTHAND REPORTERS USED. BUT IT wasn't quite right. Gridlock is an accident, an inconvenience. What happened on Capitol Hill was a strategy, and its architect was Kentucky senator Mitch McConnell.

McConnell's tactics were informed by a pair of brilliant, if somewhat evil, insights. The first was that Americans hold their president almost entirely responsible for the performance of the government as a whole. Under his direction, Republicans in Congress behaved like offensive linemen hoping to get their quarterback fired. They knew failing to do their jobs would make them look bad. But they also knew POTUS would take the hit. No matter who caused the loss, Obama's name would wind up with an *L* beside it.

McConnell's second insight was that, if he was shameless enough for long enough, he would never get the comeuppance he deserved. Some political reporters slant left, others right, but what unites them is the desire to break new stories. Kick a puppy live on camera, and everyone will cover it. Kick a puppy per day, and steadfastly refuse to apologize, and within two weeks the press moves on.

This is what happened, metaphorically at least, in the fall of 2011. Republicans voted in lockstep against funding for teachers, cops, firefighters, and laid-off construction workers. These were causes that once inspired compromise. Everyone was shocked to see lawmakers from either party oppose them. But the surprise wore off. With frightening speed, obstruction became the new normal. Reporters might as well have written about the sun rising in the east.

I found this demoralizing, and I don't think I was alone. In one set of POTUS remarks, I wrote that a frustrating thing about being president was not being allowed outside to take a walk. It was a throwaway line, a setup to an opening joke. But when I got the draft back from the Oval, President Obama's only edit had been to add to my list.

"To clear your head," he wrote. "Or jump into a car just to take a drive." At times like these, faced with an ever-growing litany of frustrations, I found myself wondering if presidential speeches were nothing more than window dressing. Really. What was the point?

And then, in December 2011, POTUS addressed the URJ.

In the weeks before President Obama spoke at the Union of Reform Judaism's biannual conference, Jarrod Bernstein, OPE's

Jewish liaison, laid out the challenge we faced. In 2008, the Jewish community had overwhelmingly supported the president. Now, however, doubt was spreading. A tall, brash New Yorker, Jarrod summarized our objective with a Yiddishism.

"They need to know he feels it in his *kishkes*," he said. (The word means "intestines," but as an idiom it translates somewhere between "heart" and "balls.")

I knew I couldn't write a better economic address than Favreau, or a better eulogy than Cody. But I didn't sit through eight years of Hebrew school for nothing. My *kishke* credentials were unmatched.

After Favs gave me the go-ahead, I spent a week putting together a draft. I talked to rabbis. I included a reference to the weekly Torah portion. Ben and Terry, the foreign-policy writers, added a section about Israel. A few minutes before the speech, as POTUS prepared to take the stage at a Maryland convention center, Jarrod asked if he needed any last-minute coaching.

"Nah," he said, "I've got this."

He was right. Addressing the crowd of five thousand, POTUS read a line about his daughter Malia, a fixture on her school's bar and bat mitzvah party circuit. There was a pause, almost imperceptibly brief. He cocked his head to one side.

"There is quite a bit of negotiation around the skirts that she wears at these bat mitzvahs," he ad-libbed. "Do you guys have these conversations as well?"

The crowd roared with laughter and applause. The president fed off the audience's enthusiasm, and the audience fed off the president's in turn. A quote from Abraham Joshua Heschel. A paean to the uniquely American success story so many Jewish immigrants had shared. A firm defense of our alliance with Israel.

OBAMA SPEAKS JEWISH, read the next day's headline in the Israeli newspaper *Haaretz*. REPUBLICANS DON'T.

That's what presidential speeches could still do. They couldn't persuade the unpersuadable. But they could remind the rest of us why we believed, renewing our faith despite frustration. Watching from the wings of the ballroom, I couldn't help but imagine my great-grandparents arriving in America, bewildered and penniless, a century before. Now here I was, helping the most powerful person on earth tell their story, and the story of so many others like them.

By the time the president delivered his satisfied thump to the podium, I had my first White House niche. I continued writing for Valerie, Bill, and the rest of the senior staff. I continued to take the POTUS speeches no one else wanted. But I also knew that if something *kishke*-related came up, I was the go-to guy.

THAT'S HOW I FOUND MYSELF STANDING IN THE WHITE HOUSE library, next to my favorite men's room, on a Friday evening in April 2012. The president's Passover message was the final item on his schedule, the only thing left between him and the weekend. And on this particular Friday, POTUS was especially in need of a break. He had just finished a trip through Asia, a weeklong odyssey of tedious summits and brutal jet lag. Now he was running two and a half hours late, which was two hours late even for him.

As we waited for the president to arrive, I made nervous small talk with the other White House staffers in the room. Jarrod was there, clutching his pocket Haggadah the way a vampire hunter might hold a cross. Luke, one of the most buoyant and upbeat members of the A/V team, sat behind a laptop to control the prompter. Hope Hall wasn't there that day, which I didn't think would matter.

It did. From the moment President Obama entered the library, the look on his face made it clear he would rather be anywhere else. Ordinarily this was when Hope would cheer him up. She would remind him he was almost through with the schedule, or recall some silly detail from earlier that day. Without her, however, POTUS remained grumpy.

"All right," he said. "One take."

He was moving through the script, the weekend so close he could taste it, when he came to a line I had cribbed from the Passover liturgy. "In every generation, there are those who have tried to destroy the Jewish people." He got halfway through it. Then he grimaced.

"Okay, wait, stop. I didn't read this on the plane. Isn't that line kind of a downer?"

"Well . . ." I stammered. But it was too late.

"I mean, this is supposed to be a party, right? What's the deal? Like, 'Everyone's out to get us, have some matzo.'"

In fact, that was exactly the deal. POTUS had summed up five thousand years of Jewish history in just eight words. Jarrod tried to explain this, paging valiantly through his Haggadah in a search for sources, but the president's patience had run out.

"Look, just—does anyone have a pen?"

I had never heard of POTUS rewriting anything on the spot before. I would never hear of it again. But there, pen in hand, he scrawled something on the draft I had printed. He began to dictate to Luke, who was typing into the old-timey word processor that controlled the prompter.

"There are those who have targeted . . ."

"There are those who what?"

Luke's struggle to keep pace only left POTUS more irritated. Standing from his chair, he stalked over to the laptop.

"Move."

With his notes in front of him, the president extended his fingers like he was about to conduct a symphony. Then, pecking deliberately, he made his edits.

In every generation, there are those who have targeted the Jewish people for harm.

I was both embarrassed and impressed. POTUS recognized that my script might, unintentionally, cause controversy. In just five minutes, he rewrote it to express the same idea, but in far more measured tones. He did so on no sleep, ravaged by jet lag, after traveling tens of thousands of miles across the globe.

This, I thought, *is why he gets to be the president.*

POTUS resumed filming, racing through his revised draft. Jarrod, Luke, and I stood perfectly still. We were almost finished—just seconds away. Then the president hit the words *happy holidays* in Hebrew (*chag sameach)* and everything went to crap.

Except for Spanish, which POTUS insisted he could pronounce flawlessly without phonetics, it was our job to spell out all foreign words. [Bon-JOOR]. [DAHN-keh SHANE]. [ah-ree-vah-DARE-chee ROW-muh]. The system worked well, unless we encountered a sound unused in English, in which case it broke down completely. Take, for example, the hard *ch* in Hebrew. This is not the soft, gentle sound from the beginning of *child.* It's the harsh, throat-clearing one from the end of *blech.* Unless you practiced it growing up, it's nearly impossible to get right.

"Thanks, and cog somatch," said POTUS. He looked at Jarrod and me, wondering if he could leave now, but we stopped him.

"It's kind of a chhhh," I said, making a noise like a lawn mower in need of repair.

"Chhh," Jarrod repeated, phlegm building in his throat. "Chhhh, chhhhhhhh." Looking deeply unhappy, the president gave it another try.

"Chawg Samayah."

"Chong Semeeyuh."

"Hagg Sommah."

POTUS was clearly finished. Jarrod and I exchanged nervous looks. "What?" President Obama said. His tone suggested that, under a different form of government, we'd have been executed by now. "What's up?"

It was in moments like these that I thought about a social studies exercise from third grade. Each of us was given an envelope. We started by filling it out the ordinary way—name, street, city—but soon expanded our scope. Country. Planet. Solar system. Galaxy. The point, I guess, was to broaden our horizons, to help realize that the universe was marvelous and vast.

Working at the White House was like performing the same exercise, but in reverse. Start by imagining the Milky Way, almost infinitely wide. Then zoom in. Our sun comes into focus first, a fiery dot in the empty sea of space. Next comes Earth, our fragile blue marble. Drawing closer we see a continent; then a country; then a small, gridlocked district ten miles square. And then, finally, we reach the very center of the universe, where a well-meaning but exasperated Protestant is being harassed by a pair of nitpicky Jews.

"I'm sorry," I said. "We need to do it one more time."

POTUS sighed deeply. But he knew how much his words mattered. He was willing to give it one more try.

"On behalf of myself and my family, hog samea," he said. He tilted his head and raised his eyebrows, the universal expression for "Good enough?" It was. The president bolted.

"Have a good weekend, everybody."

"You, too, sir."

Once I remembered how to breathe again, I returned across West Exec to call Jacqui and tell her what had happened. And it was only then, as I tried to describe the look on the president's face, that I understood how foolish I had been. Why didn't I realize he would reject such gloomy language? Why didn't I write *happy holidays* in English, instead of using a word so difficult to pronounce?

But that's the downside of White House fairy dust. In most office buildings, not screwing up is the bare minimum. At the center of the universe, not screwing up takes extraordinary, almost superhuman skill. When your business card bears the presidential seal, mistakes that otherwise wouldn't matter can snowball into catastrophe in the blink of an eye.

No wonder salmon ended up in the toilet, I thought. In the White House, it's nearly impossible to keep it out.

6

IS OBAMA TOAST?

Just north of the White House, across Pennsylvania Avenue, is a leafy, rectangular park called Lafayette Square. The place is sleepy in the early mornings, but by 10 A.M. it comes alive like the village from *Beauty and the Beast.* Yuppies in yoga pants lunge through boot-camp workouts. Chinese tourists point at squirrels the way Americans might gawk at pandas. Segway tours snake between statues, the off-balance clients struggling to keep up with their guides.

Most people only enter the square on their way to someplace more famous. For an eclectic handful, however, lingering in a park near the White House is a kind of full-time job.

Back when I passed through Lafayette Square each morning on my way to work, the park's most famous regular was also its only full-time resident. She was tiny and wrinkled, somewhere between seventy and one million years old, in a head scarf to match her baggy clothes. Her home was a tarp-covered encampment no larger than a beaver's den. She had taken up residence there in 1981 and kept watch over the White House 24-7 ever since. Her life's work was often described as a "peace vigil," but that implied a kind of flower-child optimism. The signs tacked to her shanty suggested something far more harsh.

LIVE BY THE BOMB, DIE BY THE BOMB.
Ban all nuclear weapons or have a nice Doomsday!

Because the little old lady received frequent press coverage, I knew her name was Concepcion Picciotto. The other regulars had never been profiled in the *Washington Post,* and with nothing to go by, I assigned them nicknames in my head. *Good morning, Druid!* I'd think, biking by a man wearing only a loincloth and carrying only a staff. The Druid's chest was covered in tiny, hairy curlicues. Ropey white dreadlocks dangled like hoses from his scalp. Sometimes I spotted him browsing samples at the nearby farmers' market, but more often he sat calmly on a bench, his back to the president's house.

Whistle Guy took the opposite approach. On his jet-black, beat-up bicycle, he ping-ponged up and down Pennsylvania Avenue. He never went anywhere without his World War II helmet, his camouflage jacket, and his namesake instrument in his mouth. This he blew incessantly. It was as though he were a lifeguard and the world was refusing to leave the pool.

For sheer auditory gusto, however, no one could match Preacher Man. Tall and lanky, his preferred position was just outside the White House gate. There, his head poking above the crowd of sightseers, he would spread the gospel the way a ballpark vendor hawks ice-cold beer.

"Jeeeeeee-ZUSS! Jee-EEE-EEE-EEEE-zussssss!"

How strange, I'd think, walking past him on my way to work. *That man pays my salary.*

Of course, depending on his tax bill, it's possible Preacher Man got my services for free. Even if he did file a federal return, he didn't pay me much. In 2011, the average taxpayer deposited about one-quarter of one-tenth of one penny into my account. I could have worked at the White House for ten thousand years and cost the Druid less than one of his farmers' market peaches. Still, the point remained. The American people had invested in me. Under no circumstances could I let their money go to waste.

This led to a wave of guilt whenever I watched YouTube videos at work. Was Whistle Guy really getting his .025 cents' worth?

More important, it meant I was legally banned from political activity. For most people this might seem like an oxymoron. My boss's boss was the world's best-known politician. Wasn't all activity political? But our lawyers were not most people. They came up with a strict set of rules to ensure that it was President Obama, not candidate Obama, being served by White House staff. Writing a speech burnishing POTUS's image was permitted, but directly asking people to vote for him was not. Neither was fund-raising. Except for senior staff, who were exempt, White House employees were barred from all things related to the campaign.

These prohibitions weren't new. But as 2012 began, they were newly aggravating. Six months earlier, being told I couldn't write for rallies or fund-raisers was like being told I couldn't write limericks or in Portuguese. Now, however, the reelect was heating up.

My OPE colleagues were no less frustrated. Eager to get in on the action, they began a mass exodus from Washington to the swing states. With so many employees leaving, going-away parties were held with the speed and frequency of Vegas weddings. These took place in a stately EEOB office suite that once belonged to the Secretary of War. Around 4 P.M. each Friday, we would arrive to find the conference table piled with Costco turkey pinwheels, cheap bottles of champagne, and sheet cake preordered from the Navy Mess. After a few minutes of mingling, Straut would send off the soon-to-be departed with a toast.

"Nobody embodies this movement more than Ashley," Straut might proclaim. "We don't know what we're going to do without her. But we'll sleep better knowing the reelect is in good hands." Every speech was a variation on the same theme, yet they never failed to move me. *I'm so lucky to have known them,* I'd think, dabbing my eyes with a napkin bearing the logo of Presidential Food Service. *At least they're headed to a better place.*

No official policy required us to discuss coworkers as if they were going to heaven rather than, say, Pennsylvania. It was simply the way we felt. Washington had tainted us with its original sin of bureaucracy. Now a lucky few had been deemed worthy. How pure of heart our colleagues were! How full of faith!

Yet after the last crumbs of cake had been cleared, and the cheese tray scavenged by interns, a few of us admitted there were benefits to being left behind. When I asked my coworker Emily if she planned to join the reelect, she winced.

"I don't know," she admitted. "I kind of like having job security." She said this as if it were unspeakably shameful, as if she kind of liked wearing Nazi uniforms or heaving cinder blocks from a highway overpass. I understood her guilt, but I also understood her point. In 2008, I had been gloriously unmoored. Four years later, returning to Ohio would mean giving up a once-in-a-lifetime job.

It wasn't just my career that kept me from setting sail. Jacqui and I were in a honeymoon phase. For some couples this might have meant staring ceaselessly into each other's eyes, limbs tangled like a clumsy octopus while meals went uneaten and eviction notices piled up. Luckily, neither of us was that sentimental. Instead, giddy with each other's attention, we came home tipsy and mixed the contents of my liquor cabinet into new and thrilling concoctions. "It's an Island Julep!" we declared to my roommate Amanda, pouring a bottle of crème de menthe into a saucepan of coconut rum.

A White House job, it must be said, was not always conducive to romance. There's a reason Marvin Gaye never sang about getting an e-mail from his boss's assistant and abruptly canceling dinner plans. On the other hand, there was date night at the Kennedy Center. The president and First Lady received tickets to every performance. When they didn't use them, which was almost always, their seats were distributed by lottery to White House staff. Beethoven's symphonies. Mozart's operas. *Swan Lake.* Jacqui and I relished the chance to broaden our cultural horizons.

Even more than that, we relished the free booze. I don't know who stocked the fridge in the president's box, but I suspect it was a college kid preparing for a blizzard. Mini bottles of champagne; Budweiser; M&M's; Whitman's samplers; mints; plus a few fun-size packs of almonds for nourishment. No wonder people loved the opera.

It would be unfair, I think, to say this sort of perk was corrupting. I had good, noble reasons for wanting Barack Obama to win a second term. We had an economic recovery to continue. A health care law to implement. A global threat of climate change to confront. Still, listening to the high notes of an aria, my hand on the small of Jacqui's back, a shameful thought crossed my mind.

If POTUS loses this election, who's going to pay for my Kennedy Center beer?

This danger (the election, not the beer) was one I had only recently begun to take seriously. I had always thought about Obama being defeated the way I thought about losing a finger in a garbage disposal or watching an entire *Alvin and the Chip-*

munks movie. I knew it could happen. I just couldn't imagine it happening to me. Then we reached the end of 2011, and the economy was barely improving. Our approval rating had dropped to 42 percent. Even the most basic pieces of legislation remained stalled in Congress. Maybe that garbage-disposal accident wasn't so unlikely after all.

Nor did I assume, as I had in 2008, that Barack Obama's moral clarity was certain to prevail. Back then POTUS had appeared infallible, a grand master checkmating the status quo. Now, after a year on the inside, I had learned the truth. The White House doesn't play chess. The White House plays whack-a-mole. If it wasn't an oil spill in the Gulf of Mexico, it was a military setback in Afghanistan or economic turmoil rippling across the EU. For each crisis we batted down, another popped up. With Election Day looming, the chance of another 2008-style landslide was near zero. Our only hope was to keep the moles at bay long enough to secure a second term.

Even this would be a challenge. About a year before Election Day, the *New York Times Magazine* published a cover story. The headline drained the blood from my fingers.

SO, IS OBAMA TOAST?

The article's author, Nate Silver, had successfully predicted the winner of forty-nine states in the 2008 elections. While I didn't think he was God, I would not have been surprised to learn that he and the Almighty chatted once a week. And here was his prediction for 2012: If Republicans nominated former Massachusetts governor Mitt Romney, and the economy didn't grow, POTUS had just a 17 percent chance to win. If growth maintained its postrecession average, the president's odds improved. But not by much.

Our best hope was for Republicans to nominate someone other than Romney in the primaries, but the other candidates faltered one by one. Businessman Herman Cain was alleged to have sexually harassed employees. Texas governor Rick Perry was kind of a dunce. Newt Gingrich was famously undisciplined and had unsavory business ties. Four years later, of course, these qualities would no longer be considered deal breakers, even when they were all possessed by the same person. But these were far more innocent times. When the dust settled, Romney was the last man left.

It wasn't hard to see why this made Nate Silver pessimistic about our chances. With his sculpted hair and confident jawline, Mitt Romney cut the profile of a president in a movie. You could easily imagine him punching an alien in the kisser or sending Russkies packing with a single icy stare. And his appeal ran deeper than looks. As a private citizen, Romney had built a wildly successful company. As a public servant, he had served as the Republican governor of Massachusetts, a reliably blue state. He seemed like the kind of guy who could get things done.

Still, Romney hadn't emerged from the primary unscathed. Like the fraternity pledge who acquires a tattoo saying *eggplant* in Chinese, he'd made a few mistakes in a bid for his peers' attention. He labeled himself "severely conservative." His support of "self-deportation" for immigrants eroded his bipartisan appeal.

Most helpful of all, Romney was gaffe prone. This was a greater liability than ever, since each verbal stumble now lived forever online.

Corporations are people, my friend.

I like being able to fire people who provide services to me.

Taken separately, these comments were mere hiccups. But string them together and they suggested a blind spot where ordinary Americans were concerned, the BMW zooming down the shoulder of the highway, wondering why the rule-following

Fords and Hondas are upset. POTUS was still less popular than we would have liked. Romney led in plenty of polls. But between our opponent's emerging flaws and the fact that a large chunk of McCain's base had died of old age, Obama wasn't toast just yet.

"Do we have a shot?" friends asked.

"It'll be close," I'd say, toeing the cautiously optimistic party line. "But I'd rather be us than them."

Us and them. Obama's election was supposed to end that kind of thinking. Watching the inauguration on a barroom screen, I was certain a new era had arrived. We would heal old wounds. We would find new ways to work together. Sure, we would have our disagreements, but always with the knowledge that we belonged to the same team.

Instead, the opposite had occurred. In just a few years, dividing lines had rippled across America like cracks in a car windshield. It wasn't just conservative versus liberal. Each time I biked to work I passed McPherson Square, the local headquarters of the new movement known as Occupy Wall Street. Pedaling by their tarp village in my navy-blue suit, I could feel their antiestablishment loathing seeping through the fabric. *But I'm a Democrat!* pleaded the voice inside my head. This led to an imaginary argument that ended, inevitably, in rage. *Hey, pal, I may not be a white guy with dreadlocks, but I believe in justice, too!*

At other times, however, I had to concede the occupiers' point. This was especially true in January 2012, when I worked on POTUS's speech before something called the Alfalfa Club.

Put ten thousand protestors in a tent camp with ten thousand typewriters, and they would never dream up anything as offensive as the Alfalfa. Founded in 1913, the club's stated mission is to bring together America's most powerful people for no good reason at all. The sole guiding principles are (a) an excess of food and (b) an excess of alcohol. It's the grown-up equivalent of the fridge at the Kennedy Center, minus the symphony and ballet.

The club meets just once a year, on the third Saturday of January, in the ballroom of a Washington hotel. There, decked out in tuxedos and gowns, two hundred Alfalfans dine on lobster and steak while roasting each other gently. (Jokes about money are acceptable, for example. Jokes about power are not.) Along with "nominating" a presidential candidate to give a tongue-in-cheek acceptance speech, the club inducts new members. These are grown men and women who, for the term of their initiation, willingly agree to be referred to as "sprouts." Ridiculous, I know.

But before you dismiss the Alfalfa entirely, consider who the list of former sprouts includes: Henry Kissinger. Warren Buffett. Alan Greenspan. Sandra Day O'Connor. Mike Bloomberg. Neil Armstrong. Newt Gingrich. Chuck Schumer. Madeleine Albright. Colin Powell. Steve Forbes. Bill Clinton. Jeb Bush. George W. Bush. George H. W. Bush. Others have names you wouldn't recognize, but represent organizations you would. Marriott. Procter & Gamble. The army. The air force. General Motors. Goldman Sachs. All have a seat at the table.

I doubt many Alfalfans see their club as a tool for protecting the status quo, but that's the way it's worked out. The organization barred African Americans until a decade after the Civil Rights Act. Doors were closed to women until 1994. Even the banquet's date is retrograde: it honors the birthday of Robert E. Lee.

It's hardly surprising that America's first black president found the Alfalfa less than wonderful. In 2009, Barack Obama spoke there, just as every president had before him, but he didn't exactly pay his respects. "Look at the person sitting on your left," he said. "Now look at the person sitting on your right. None of you have my e-mail address."

Nor did POTUS let the Robert E. Lee thing slide. Instead, he thrust it into the spotlight: "If he were here with us tonight, the general would be 202 years old. And very confused."

By normal standards this was nothing. At the Alfalfa, it was the rough equivalent of insulting everyone's mother and farting loudly into the mic. When POTUS left, I imagine he vowed never to return.

Yet in 2012, he returned. Plenty of the speechwriters were livid. The club was the embodiment of everything we had promised to change. Was it really necessary to flatter these people, just because they were powerful and rich?

In a word, yes. In fact, thanks to the Supreme Court, the rich were more powerful than ever. In 2010, the court's five conservative justices gutted America's campaign finance laws in the decision known as *Citizens United*. With no more limits to the number of attack ads they could purchase, campaigns had become another hobby for the ultrawealthy. Tired of breeding racehorses or bidding on rare wines at auction? Buy a candidate instead!

I should make it clear that no one explicitly laid out a strategy regarding the dinner. I never asked point-blank if we hoped to charm billionaires into spending their billions on something other than Mitt Romney's campaign. That said, I knew it couldn't hurt. Hoping to mollify the one-percenters in the audience, I kept the script embarrassingly tame.

I've got about forty-five more minutes on the State of the Union that I'd like to deliver tonight.
I am eager to work with members of Congress to be entertaining tonight. But if Congress is unwilling to cooperate, I will be funny without them.

Even for a politician, this was weak. But it apparently struck the right tone. POTUS barely edited the speech. A few days later, as a reward for a job well done, Favs invited me to tag along to a speechwriting-team meeting with the president.

I had not set foot in the Oval Office since my performance of the *Golden Girls* theme song. On that occasion, President Obama remained behind his desk. For larger gatherings like this one, however, he crossed the room to a brown leather armchair, and the rest of us filled the two beige sofas on either side. Between the sofas was a coffee table. On the coffee table sat a bowl, which under George W. Bush had contained candy but under Obama was full of apples instead. Hence the ultimate Oval Office power move: grab an apple at the end of a meeting, polish it on your suit, and take a casual chomp on your way out the door.

I would have sooner stuck my finger in an electrical socket. Desperate not to call attention to myself, I took the seat farthest away and kept my eyes glued to my laptop. I allowed myself just one indulgence: a quick peek at the Emancipation Proclamation.

That's right, buddy. Look who's still here.

It was only at the very end of the meeting, as we rose from the surprisingly comfy couches, that Favs brought up the Alfalfa dinner. The right-wing radio host Laura Ingraham had been in the audience, and she was struck by the president's poise. "She was talking about it this morning," Favs told POTUS. "She said, 'I don't know if Mitt Romney can beat him.'"

By this point, President Obama was showing us the door. But when he heard Favs's story he paused briefly and puffed up. It was like watching a boxer at a weigh-in.

"Well," he said, "Mitt Romney can't beat me."

If chest bumping had been permitted in the Oval, we would have gone for it. Instead we did the next best thing, laughing with outsize confidence as we strutted from the room.

The incident left me with no grand illusions regarding comedy. I didn't think a few one-liners could create new jobs or push our approval above 50 percent. Still, I was proud. Without violating the ban on political activity, I had helped POTUS send a warning shot across Romney's bow. Maybe I even discouraged an anti-Obama SuperPAC. The point was that I had played some small role in the biggest political battle of my lifetime. And I hoped that when the 2012 Correspondents' Dinner rolled around, I'd have the chance to do it again.

WHEN I WORKED AT THE WHITE HOUSE, PEOPLE OFTEN ASKED ME IF I used jokes to advance the president's agenda. I always said no, by which I always meant yes. That's not to say each punch line was poll tested. I simply felt that if the leader of the free world was required to host a comedy night, it ought to be worth his time.

Under Jon Lovett's supervision the previous year, POTUS's assault on Trump had more than met this standard. The conventional wisdom, which I wholeheartedly subscribed to, was that Obama had "destroyed" the birther king. He had "demolished" him. But in late 2011, Lovett moved to Hollywood to write sitcoms. I became, by default, the White House's token funny person. As I prepared to run the joke-writing process for the first time, destroying and demolishing Romney seemed like obvious goals.

This was easier said than done. Back in 2011, Trump was the rare breed of public figure loathed by Democrats and Republicans alike. Romney was different. He had friends. Step even an inch over the line, and they would complain to reporters, who would milk the ensuing controversy for days.

With a frontal assault out of the question, the best we could hope for was a series of bank shots. By laughing at his own expense, POTUS could appear both confident and humble. By injecting arguments directly into setups and punch lines, he could bypass the media back-and-forth. Perhaps most important, by joking about controversy, he could diffuse it. *If the president's willing to laugh at something, how bad can it really be?*

By the time I began writing jokes, three weeks before the dinner, there was plenty of controversy to diffuse. In March, a hot mic caught POTUS telling Russian president Dmitry Medvedev he'd have "flexibility" after the election. Policy-wise, this was reasonable. Stripped from context, with the frisson of excitement that comes from eavesdropping on world leaders, it looked bad.

So did the debacle unfolding at the Government Services Administration. The agency's sole purpose was to spend taxpayer money wisely. Yet it had recently shelled out more than eight hundred thousand dollars for a Las Vegas conference featuring a mind reader and a clown. And GSA staff weren't the only federal employees whose entertainment choices had gotten them in trouble. On a recent trip to Colombia, several Secret Service agents had been caught soliciting prostitutes when one of them skipped out on the bill.

Then there were the scandals involving dogs. I know that sounds absurd. It *is* absurd. But the fact remains: in April 2012, two canine controversies were major political news.

The first dated back to 1983, when a young Mitt Romney drove his family to a vacation home in Canada. This was unremarkable. What *was* remarkable is that, the car overstuffed with bags and children, he had transported his family's Irish setter in a carrier strapped to the roof. Hoping to contain the fallout, Romneyites dug up a scandalous story of their own. In his autobiography Obama admitted that, as a six-year-old in Indonesia, he had eaten dog meat.

The whole thing was stupid. Neither anecdote said much about the president each man would make. Yet political commentators couldn't get enough of these stories. Was it worse to mistreat a dog as an adult or ingest one as a child? In 2012, somewhat astonishingly, plenty of political reporters felt this question was worthy of their time. And it was my mission to leave no controversy, real or nonsensical, unaddressed.

To that end, I compiled a long list of topics and sent them out to what amounted to our writers' room. David Axelrod and Jon Lovett sent in jokes from the growing Obama diaspora. Other one-liners came from Jeff Nussbaum and his West Wing Writers team. More than a half dozen contributors came from entertainment rather than politics. Nell Scovell created *Sabrina the Teenage Witch*. Kevin Bleyer was a producer on *The Daily Show*. Nina Pedrad wrote for *30 Rock* and *New Girl*. Judd Apatow was (and is) the leading comedy mogul of his generation, behind everything from *Girls* to *Knocked Up* to *Freaks and Geeks*.

Perhaps our friends in Hollywood knew how to crank out an endless series of amazing jokes. I certainly didn't. For me, the secret to writing one funny line was to write about twenty-five awful ones first. Most evenings I would comb through the day's rubble and sigh. But after picking the diamonds from the rough, and combining them with material coming from outside the building, a monologue began to take shape.

A few days before the dinner, Favs and I went to the Oval to present about forty of our favorite lines. Together with David Plouffe, the president's senior advisor, and Jay Carney, the press secretary, we sat on the couches while POTUS read out loud. Each time he laughed, I made a mental note. Each time he didn't, I had a mental breakdown.

Then he reached my first dog joke and my heart stopped beating entirely. It relied on an obscure reference to one of Sarah Palin's lines from the 2008 campaign. It also involved eating man's best friend. As POTUS read off the page, I wondered if we hadn't made a mistake putting it before him.

"Sarah Palin's getting back in the game, guest hosting on the *Today* show. Which reminds me of an old saying: What's the difference between a hockey mom and a pit bull?"

He paused, just slightly.

"A pit bull is delicious."

We all fell silent. Was POTUS confused? Was he disgusted? Then, to my enormous relief, he grinned.

"That's pretty good," he said, chuckling. "A pit bull is delicious. I like that." He smiled again, mulling things over. "I'm just letting you know, I might add something here. 'Maybe a little soy sauce.' Something like that." He made a shaking gesture with his hand, sprinkling imaginary seasoning onto his canine meal.

As my heart resumed beating, President Obama finished reading the jokes, culling about a dozen or so from our list. Then he ushered us out of the Oval, shaking his head in mock disbelief. As I left the room, I heard him chuckle. When he spoke, it was with the same officious tone he used when acknowledging a congressman in the crowd.

"A pit bull," he proclaimed, "is delicious."

THE DAY BEFORE THE DINNER, WE GATHERED IN THE CABINET ROOM to record a short audio piece making fun of the hot mic with the Russian president. POTUS sat near the door to the Outer Oval, directly beneath a portrait of Harry Truman. I sat next to him, cast as "White House Aide."

I had only one line: *Mr. President, they're ready for you.* But with Barack Obama sitting to my right, I felt the tightness in my chest that had ruined my high school dreams of stardom.

"Mr. President they're! Ready-FOR you."

Ignoring my botched delivery, POTUS recorded his lines.

"I'm the president of the United States, and I'm opening for Jimmy Kimmel?"

"Right now I'm like a five on the Just for Men scale. I think I could go to six and nobody would notice."

"I could really use a cigarette."

POTUS was a better actor than I was. Still, his timing was off. He emphasized some of the wrong words. He paused in awkward places. We were allotted only ten minutes, enough for just two recordings, and I began to wonder what would happen if his delivery didn't improve.

I certainly wasn't helping. As we began our second and final take, I flubbed my line even worse than before.

"MR. PRESIDENTthey'rereadyfor . . . you?"

This made what happened next all the more remarkable. I hadn't taken my eyes off President Obama. I knew for a fact he had not practiced. And yet the difference between his first and second read-throughs was the difference between a guy puffing through kickboxing class and Jean-Claude Van Damme. He took beats at just the right moments. He hit the precise words to sell each punch line best. His tone was the perfect blend of annoyance and self-regard. It was as if he'd spent a full day rehearsing. It was that much better.

I'd often heard senior staff describe President Obama as the smartest guy in the room, but only now did I realize what they meant. He didn't speak seven languages or know the Latin names of species or multiply large numbers in his head. What he did, more quickly than anyone, was strip away complicated issues to their essence and make the most of the information obtained. No one was better at getting to the point.

Jon Lovett returned from Los Angeles for the dinner that year, and on the morning of April 29, he, Favs, and I did one final run-through in the Oval. By now the script was nearly final. In fact, we had only one new line to run by President Obama. During a recent speech, Vice President Biden had remarked that POTUS had a "big stick." He was referring to foreign policy, but his hand gesture was, to put it mildly, undiplomatic. Jeff Nussbaum pounced.

"Let's just put it this way," read his joke. "Dreams aren't the only thing I got from my father."

POTUS laughed so loudly that I secretly hoped he would add the line to the script. But this was an election year; a presidential dick joke was a bridge too far. And so, with nothing more to add, we were finished. Favs and Lovett went to a fancy brunch party. POTUS went to play golf. I went home to nap.

I ATTENDED THAT YEAR'S CORRESPONDENTS' DINNER AS A GUEST OF the Agence France-Presse. It's a wonderful organization, but they're hardly the cool kids of the media world. Where the *Time* and *Bloomberg* tables overflowed with celebrities, the seat to my right was occupied by a well-regarded Irish novelist. His face seemed sculpted from putty. His eyes glittered beneath thick black brows. He was not exactly a movie star.

In fairness, however, I was not exactly a Washington insider. Perhaps that's why we made quite a team. As we passed around the basket of rolls, my novelist explained his theory of *Hamlet*. I nodded, genuinely absorbed. Even better, by the time the salad course was finished, he was on his fourth glass of wine and his stage whisper had become a full-throated scream. At one point during the evening, Jake Tapper, then a reporter with ABC News, stood to receive a prize from the Correspondents' Association. My tablemate snorted.

"Really?" he shouted. "Now they're givin' feckin' awards to their feckin' selves?"

The wine continued disappearing, and by the time President Obama took the stage, I had a one-man cheering section. After the audio making fun of his hot mic moment, POTUS began with a joke about the bin Laden raid.

"Last year at this time—in fact, on this very weekend—we finally delivered justice to one of the world's most notorious individuals." The giant screens on either side of the podium displayed a picture of a sneering Donald Trump.

"Did you write that?" my novelist mouthed. When I nodded, he gave me two enthusiastic thumbs-up.

My mouth was dry. I clutched my chair nervously. But even in my deer-in-headlights state, it was amazing how quickly three weeks of work flew by. Before I knew it, the president was delivering his line about eating pit bulls, adding imaginary soy sauce as promised. We played a short video from a SuperPAC called Dogs for Romney, defending the right of canines everywhere to ride on the roof. For the last line of the night, the president returned to damage control.

"I had a lot more material prepared, but I have to get the Secret Service home in time for their new curfew."

Just like that, it was over. POTUS sat down, Jimmy Kimmel told fifteen minutes of jokes, and the entire Hilton ballroom began streaming out the door. I was trying to make sense of it all when I saw someone from the corner of my eye.

"Oh my gosh. That's Diane Keaton!" I don't usually gape at celebrities, but this was one of my all-time favorite actresses. And now she was headed in our direction, dressed in a bowler hat, jacket, and tie.

My cheering section didn't waste an instant. "Let's introduce ourselves!" My novelist took off in pursuit like a cheetah in a nature documentary, only drunk. Diane Keaton saw him coming. She tried to flee. But it was hopeless. Bottlenecking her between two tables, my new friend pounced, cheerfully putting an arm around the actress's waist. "Look at what I bought at the train station!" he shouted, as though this were an acceptable form of greeting. Then he reached into his pants pocket and produced a cardboard disposable camera. It was the kind I remembered from summer camp, the kind where you advance the film by rotating a plastic wheel.

"Don't we want to see if it works?" my novelist asked.

"No, that's okay," said Diane Keaton.

"Come on, we have to test it. It'll be fun!"

"Well, really, I . . ."

"All right then." He grabbed me by the arm, pulled me close, and took a decidedly old-fashioned selfie. "Oh good, it works!" he announced.

"That was . . . impressive," I said, as he released his trophy back into the wild. It was a fittingly surreal end to the most grueling three weeks of my life.

I WALKED INTO THE OFFICE ON MONDAY ASSUMING THAT, WITH THE dinner over, I would simply pick up where I left off. I would write speeches for senior staff, and occasionally for POTUS, always on the lookout for remarks involving jokes, Jews, or some combination of the two. If only it were that easy. Something was missing. Was it the joy of writing comedy? The chance to eavesdrop on Arianna Huffington and George Clooney? The fact that my Diane Keaton picture never arrived in the mail?

But as the weeks rolled on, I realized it wasn't a lack of glamour that was bothering me. Instead, I kept thinking back to a line Valerie liked to include in her commencements: "Put yourself in the path of lightning." For just one night, a seventeen-minute comedy monologue was the center of political attention. It was the place to address controversies, to take shots at opponents, to project confidence to the public we served.

Now, however, lightning was once again striking the campaign trail. More and more speeches—for both the president and senior staff—were the ones I could not legally write. I kind of liked having job security. I kind of loved drinking Kennedy

Center beer. But nothing was as intoxicating as being part of the action.

Not long after the dinner, I asked Favs if I could leave the White House for the campaign. He agreed, but proposed a plan that kept me in Washington: I would work on political speeches for POTUS, but from the Democratic National Committee in D.C.

Which is how I found myself, a few weeks later, standing beside a conference table covered in turkey pinwheels and cheap champagne. Straut said something generous. Coworkers wrapped leftovers in paper napkins. I turned in my blue badge and BlackBerry. Just like that, I was no longer a government employee.

"Jeeeeeee-ZUSS! Jee-EEE-EEE-EEEE-zussssss!"

Leaving the building, I passed by Preacher Man. Whistle Guy's screeching echoed down the street. Walking to the bus stop I noticed the Druid, stoic as always on his bench. It had only been a year, yet somehow I'd grown used to having them around.

How strange, I thought. *These people no longer pay my salary.*

Would I be back a few months from now? Was I leaving for good? That was up to the voters to decide.

7

GOING EASTWOOD

The Republican National Committee headquarters, perched conveniently on Capitol Hill, is everything an office building should be. The four-story façade is as white and well-maintained as the people who work there. Gray accents provide a restrained pop of color. Tasteful moldings lend class. Imagine if a bank branch mated with a country club and raised a perfect child. That's the Republican National Committee. It's a lovely place.

Compare this to the Democrats. As far as I can tell, the DNC headquarters was designed by someone who had never seen a building before. Mud-colored concrete follows a blueprint of random angles and arbitrary curves. Gray metal awnings punctuate walls the way a cat on a keyboard punctuates a sentence. Rainwater pools in ill-conceived balconies, trickling to the sidewalk and leaving rusty streaks behind. My freshman year of college I wore a fedora to class in the sincere belief it would establish me as both stylish and smart. The outside of the

Democratic National Committee is the architectural equivalent of that hat.

And yet, remarkably, the inside is worse. The humorist Will Rogers once declared, "I belong to no organized political party. I am a Democrat." He meant it as a joke, but whoever came up with the DNC floor plan clearly took it as a motto. For example, why insert the Democratic Congressional Campaign Committee, by law a separate entity from National Committee, right in the middle of the second floor?

That might not sound like much of a problem. It was a nightmare. My new desk was also on the second floor of the DNC and should have been a short walk from the elevator. But to avoid even a hint of unlawful collusion, I was barred from entering the Congressional Committee office at any time. Legally speaking, their carpets were made of molten lava. This meant that each morning I took the following route. First, I rode the elevator to the third floor. Then I walked the entire length of the building, past the reception desk, the break room, and the fund-raising offices. Next, I descended a fluorescent-lit flight of neglected concrete steps. Having reached the second floor, I then walked halfway back across the building, *in the exact opposite direction*, until I reached my cubicle. This daily trek didn't exactly make me Lawrence of Arabia. But neither did it suggest a well-oiled machine.

I suppose I could have followed the example of my intern, an eager poli-sci student named Devlin. Devlin showed up early. He never once complained about the labyrinthine commute. But such cheerful blindness to our surroundings proved impossible for me to summon. *Of course he's excited about everything,* I thought. *He's twenty-one.* I was now twenty-five, and grizzled by a year of government service. Even small signs of dysfunction were exhausting.

Before long, I began working from home. Presidential campaigns are often described in military terms—battleground states, foot soldiers, air cover—but in 2012, my war room was a French bakery called Le Caprice. Each morning I'd tell myself I was only buying coffee. Then a poll would have Romney up in Colorado, or a Senate Democrat would say something dumb.

Oh, well, I'd think. *Better have a chocolate croissant.*

Around 10:30 A.M., after my first almond cookie but before my goat cheese and prosciutto sandwich, I'd get to work. This was less thrilling than my new job title implied. "Speechwriter to the President" suggested access and influence. In reality I was a kind of rhetorical handyman, keeping our stump speech up to code. Changing *Iowa* to *Ohio* and vice versa. Tracking down numbers for Favs and Cody. Turning talking points into remarks and remarks back into talking points. Compared to 2008, my job was way more important. It was also way less fun.

I wasn't the only one making this trade-off. There's a saying in Washington: "You campaign in poetry and govern in prose." But the reelect was nothing if not prosaic. Obama's soaring rhetoric was supplemented with a careful defense of his record. Twenty-thousand-person crowds at sports arenas gave way to two-thousand-person crowds at high school gyms. Decisions in the field, once made on a kind of cowboy instinct, were increasingly driven by data analytics. It drove my organizer friends nuts.

And yet sitting in Le Caprice, my lap dusted in cookie crumbs, I didn't really mind. After all, America was not exactly clamoring for an encore of 2008. Back then, the country was falling in love. Imperfections were easy to overlook. Now, nearly half a decade into a relationship with Obama, voters weren't so forgiving. They still liked their president. They believed his heart was in the right place. But was he perfect? Metaphorically speaking, the nation had noticed that its leader left dishes in the sink.

So, no, from a political perspective, a prosaic campaign didn't worry me. From a personal perspective, however, the dangers were harder to ignore. Weeks on the Le Caprice diet had produced a noticeable softness around my middle. Devlin's reports from headquarters were growing lonelier by the day. Worst of all, I was losing my sense of connection to the people I theoretically served. In Washington you never stop hearing about the details of policy, but you rarely see its effects. Imagine if no one who worked in Hollywood went to the movies or watched TV. It's a strange condition. It eats away at you.

And if you're not careful, it can dissolve whatever brought you to our nation's capital in the first place. By now, I knew my share of go-getters who moved to Washington to change the world but had since become hollowed out. They held fancy-sounding jobs but took hour-long naps under their desks. They set up conference calls to discuss other, earlier conference calls. They could sniff out an open bar like a pig hunting truffles, but were physically incapable of paying attention to anyone who couldn't help their careers.

I wondered what transformed these onetime idealists into connoisseurs of networking and sloth. Then I wiped chocolate from my upper lip and ignored Devlin's latest e-mail. Maybe I already knew.

In August, as my concern over moral flabitude neared a breaking point, my old boss Jeff Nussbaum appeared with an offer. The Democratic National Convention was just one month away, and along with the marquee speeches—POTUS, FLOTUS, Biden—there would be dozens of remarks seen by almost no one outside the hall. Jeff was responsible for making sure these second-tier speeches went smoothly. Could I take a ten-day leave of absence from my DNC job, come to Charlotte, North Carolina, and help out?

On one hand, nothing about the work sounded fancy. I wouldn't be writing for POTUS, not even in a handyman role. On the other hand, this was a chance to leave Washington and visit a bona fide swing state. I told Jeff I needed time to think it over. But really, I couldn't wait to go.

POLITICAL CONVENTIONS WERE ONCE INFUSED WITH DRAMA. EVERY four years the nation's cronies gathered in back rooms to chomp cigars, shout "Huzzah!" for their preferred candidates, and emerge, days later, with a consensus nominee. But that was decades ago. Today conventions are music festivals for people who flip out over Senator Claire McCaskill but have absolutely no idea what a Skrillex is.

TV executives have noticed that this is not for everybody. On the big three networks, each of our four nights in Charlotte was only carried live from 10 to 11 P.M. But inside the hall, the parade of speakers began hours earlier. The governor of Connecticut. The CEO of Costco. The president of the AFL-CIO. There were more than a hundred speakers in all, each one with a few precious moments on the party's biggest stage.

Hence Jeff's real responsibility: Keep these people in line. Don't let them say something embarrassing. Don't let them bore the audience to tears.

Above all, don't let them go over their allotted time. If you ever are asked to speak at a political convention, I have no doubt you will find yourself thinking you deserve an extra minute. I am here to tell you something. You don't. It doesn't matter how important you are outside the arena. Unless you're being nominated for president, you're singing backup, even when you take the stage.

To help him corral his herd of VIPs, Jeff had recruited the speechwriting version of the Suicide Squad. There was John "J. P." Pollack, winner of the 1995 O. Henry Pun-Off World Championship. Alexandra Veitch, whose pearls and cardigans hid the rhetorical equivalent of a shiv up her sleeve. Sarada Peri, who did hard time at Teach for America and the Kennedy School of Government. Andy Barr, the surprisingly upbeat grouch (or surprisingly grouchy optimist) who ran comms for Al Franken's first Senate campaign. And then there was me, glowing faintly from White House fairy dust, still terrified by the idea of meeting with roommates to discuss a chore chart but increasingly comfortable in a suit.

Each morning we trudged into the bowels of the Time Warner Cable Arena like the dwarfs in *Snow White*. Ordinarily home to the Charlotte Hornets, the 780,000-square-foot building had been transformed into a temporary home for the campaign. For some departments, working out of luxury boxes or spacious concession areas, this was a major upgrade.

For speechwriting, it was not. We were handed the referees' locker room, a space designed for three people to quickly change clothes before and after games. There were fifteen of us, working sixteen-hour days. We were crammed together like supermarket lobsters. It was as far from Le Caprice as I could get.

And yet I loved it. Something about our combination clicked. Besides, the election was perilously close. In Charlotte, everything we'd fought for over four long years would be defended over just four days. Who cared if I was literally rubbing shoulders with my coworkers? Who cared if a flimsy plastic curtain was the only thing separating our office from the bathroom on the other side? Lightning was striking left and right.

We were free to leave our basement office—lone speechwriters would occasionally escape for snacks or coffee—but only once did we venture to the surface as a group. About a week before the convention started, we walked five blocks from the arena to a downtown office tower. In a dusty conference room, we gathered around a gray plastic conference table. Then, like a bearded, bushy-eyebrowed guardian angel, Joel Benenson appeared.

When it comes to clothing, political strategists have two

options: you can dress smart, or you can dress a little schlubby because everyone already knows you're smart. Joel was firmly in the latter camp. His suits were slightly baggy. His shoes were more comfortable than stylish. No one cared. Our campaign's chief pollster was quite possibly the world's leading expert on America's middle class. Each night his surveys went into the field like an army of tiny therapists. "So, the recession is finally over. How does that make you *feel*?"

Now, standing before our suicide squad, Joel presented his findings: President Obama's narrow lead was even more fragile than it appeared. A small but significant percentage of voters remained undecided. They held the president's fate, and ours, in their hands.

At first glance this might seem strange. We had just been through four years of nonstop partisan conflict. Who could possibly be undecided at this point? And it's true: for most of us, a vote for president is as fixed and certain as the answer to a password recovery question. *What was your first pet's name? What was your high school mascot? Which candidate do you support?*

For a sizable number of Americans, however, presidential preference is surprisingly malleable. *Would you cheat on your husband? Would you sleep with George Clooney?* In politics, sometimes the answer depends entirely on the questions being asked.

This was the essence of Joel's research. When the election was framed as a performance review—Has Obama met expectations?—we did poorly. The economy was growing more slowly than most voters would like. Our best hope was not to change minds, but to change the subject. *If you had to choose between Obama and Romney, who would you rather have as your boss?* When the election was seen as a job interview—a choice between two potential hires—voters chose Obama every time.

To make this choice as clear as possible, Joel field-tested a smorgasbord of messages. At the conference table in Charlotte, he presented us with the winner:

"President Obama understands that the economy grows not from the top down, but from the middle out and the bottom up."

These fourteen words were drilled into our heads like the Lord's Prayer or the rules of Fight Club. *Not from the top down, but from the middle out and the bottom up. Not from the top down, but from the middle out and the bottom up.* We dutifully inserted the line in speech after speech. This is what's known as "message discipline," but the existence of a technical term doesn't make it feel less silly. Imagine going to a cocktail party full of strangers. Now imagine being told that, in every conversation, you're expected to say that Shake Shack owes its success less to savvy marketing than to its proprietary blend of meats. It's weird.

The only thing more bizarre than message discipline, it turns out, is a lack of it. As we were preparing for our convention, Republicans were finishing theirs, and on August 30 we huddled in our hotel lobby to watch Mitt Romney's big speech. All week long, rumors had buzzed. Republicans held an ace up their sleeve. Was it a well-known Democrat? A prominent general?

When the surprise guest was revealed to be Clint Eastwood, the worried knots in our stomachs untangled into confusion. Allotted five minutes of stage time, the actor-director rambled for twelve. Even stranger, he spent the majority of his time pretending a nearby chair was the president of the United States. GOP convention-goers thoroughly enjoyed the performance. But outside the hall, thirty million Americans had tuned in hoping to see a potential president, and were forced to watch an octogenarian insult furniture instead.

In our hotel lobby, we exchanged joyous, disbelieving looks. Thanks to a poorly timed, profoundly odd spectacle, Romney's moment in the spotlight had been completely undercut. That gave us an opportunity. It also gave us a mission. Make sure no one goes Eastwood in Charlotte.

ON MONDAY, JUST HOURS BEFORE OUR CONVENTION BEGAN, I received a surprise assignment. It came from Erik Smith, the longtime Democratic operative in charge of the program.

"Big news," he said. "Natalie Portman, Scarlett Johansson, and Kerry Washington are speaking on Thursday. Want to write the script?" His offer came with implied wink and nudge, as if he were appointing me bikini inspector.

"Sure," I said. But I didn't think much about it. I simply added the actresses to the long list of speakers I'd been assigned. Most were no strangers to the soapbox. Former Virginia governor Tim Kaine. Education Secretary Arne Duncan. Democratic Party chair Debbie Wasserman Schultz. A Miami talk-show host named Cristina Saralegui. For them, appearing in public was routine.

For a few others, however, speaking at a convention was as far from normal as you could get. These men and women were not politicians. They were students and parents and factory workers who had been helped by the president's policies. They led lives viewers at home could relate to. Now they were being asked to share their stories, in front of twenty thousand people, in two minutes or less. It was one of these stories that occupied the bulk of my focus as our convention began.

I first learned about Zoe Lihn from a short film produced by the campaign. The video began with a two-year-old girl giving an eager kiss to a purple stuffed sheep. Then her mother faced the camera. "My name is Stacey Lihn, I'm from Phoenix, Arizona, and the Affordable Care Act is saving my daughter's life."

As the video continued, Stacey explained that Zoe was born with a congenital heart defect. At just fifteen hours old, she endured her first open-heart surgery. A second surgery followed four months later. Here, Zoe interrupted the video to proudly offer her dad a piece of fruit.

"Apple!"

"That's an orange."

"Oh."

It was adorable. It was also heartbreaking. When Zoe was born, insurance companies were allowed to place limits on the total amount of coverage one person could receive. At just six months old, Zoe was halfway to her lifetime cap.

Then came the Affordable Care Act. Thanks to Obamacare, lifetime limits on coverage were prohibited. Stacey and her husband, Caleb, could once again afford their daughter's care. But Zoe needed at least one more surgery to repair her heart, and it was scheduled for 2013. If the law were repealed and the lifetime caps reinstated, the Lihns would be helpless. For Zoe, the 2012 election could mean life or death.

After three years as a speechwriter, I had grown slightly numb to gut-wrenching stories, the same way medical students grow accustomed to the sight of blood. But the Lihns were different. Stacey's authenticity was almost impossible to find in Washington. Her talent for connecting the personal and political was almost impossible to find anywhere else. I was determined not to let her down.

In an ideal world, Stacey and I would have met in person on multiple occasions, over a period of weeks, to carefully prepare her remarks. In the real world, we had twenty minutes on the phone. With so little time, "speechwriting" was not much more than glorified transcription. I wrote down everything Stacey said. I cut all but the 260 most memorable words. Then e-mailed them to her so she could rehearse.

By the time I finally met Stacey face-to-face, she was just hours from taking the stage. From the campaign video, I knew she wasn't tall or broad shouldered. But when I saw her sitting outside one of the prep rooms, I was struck by just how tiny she was. Our speech coach, a woman who dressed like a high school principal and specialized in fine-tuning delivery, was not exactly a giantess herself. Yet she towered over Stacey as they shook hands. Then Stacey walked to the practice podium.

"Governor Romney says that people like me were most excited about President Obama the day we voted for him. But that's not true. Not even close."

At times, her voice quavered. Her eyes filled with tears as she struggled to keep emotions in check. But behind her fear and frustration I could see unshakable courage, the Pomeranian that thinks it's a pit bull.

When Stacey reached the end of her remarks, the speech coach administered instructions. Pause here. Breathe there. Don't be afraid to speak over audience clapping; the sound of applause fades faster on TV. Stacey did one more run-through and then prepared to leave the room. But the speech coach wouldn't let her. I figured she was about to suggest another word to emphasize, or offer a tip on modulating one's voice.

"I wanted to let you know something," the coach said. "My daughter has a congenital heart defect, too." Suddenly, we weren't operatives anymore. Everyone was crying.

On the convention's first evening, Michelle Obama gave one of the best addresses I had ever heard. San Antonio mayor Julian Castro delivered an impressive keynote. In the referees' locker room, we breathed easier—our first day in Charlotte was Eastwood free.

But for me, one highlight stood above the rest. At 8:42 P.M., Stacey Lihn took the stage in a purple dress, with just the tiniest bit of terror in her eyes. Caleb followed, holding Zoe in his right arm while Zoe's big sister, Emmy, trailed behind. Voice trembling at first, but growing stronger with each sentence, Stacey spoke about the relief the Affordable Care Act had brought her family. She spoke about the possibility that Zoe might need a heart transplant.

"When you have a sick child, it's always in the back of your mind, and sometimes in the front of your mind. On top of that, to worry that people let an insurance company take away her health care, just because of politics?"

In the three years and eight months since Barack Obama had taken office, important laws had gone unpassed. The Tea Party reanimated America's darkest instincts. The new House majority made partisanship a way of life. No wonder that, to most Americans, *politics* had become a dirty word.

But here was this mighty speck of a woman. She hadn't sought the spotlight. With so many challenges facing her family, no one would have blamed her for leaving the big picture to someone else. Yet she saw a connection between the man who held the nation's highest office and a two-year-old girl who held a stuffed purple sheep. She was willing to open her heart to twenty thousand strangers in Charlotte because she knew her fight was not hers alone. That was politics, too.

IN WASHINGTON, THE WAY YOU REFER TO IMPORTANT PEOPLE IS governed by a web of unwritten rules. If you and a VIP have no real relationship, you do not fake it. Instead, you use their last name and title, even when talking with friends.

"Governor Kaine sent some edits."

"I just had speech prep with Secretary Duncan. He was really down to earth!"

If you're familiar with the boss, shorthand becomes acceptable. But for those outside the inner circle—a staff assistant, for example—first names remain taboo. Initials are preferred. "CVH" for Congressman Chris Van Hollen. "DWS" for Debbie

Wasserman Schultz. Even senior advisors shy away from first names. They signal familiarity by naming the office instead.

"Heads up, the governor has a few edits."

"You'll like the secretary. He's really down to earth."

You are free to use the first name of a powerful person only when the following two conditions are met. One, you must have a genuine working relationship. Two, the VIP in question must still, on some level, be staff. I could write *Valerie* in e-mails. Back when I was avoiding *S* sounds in his speeches, I could refer to the chief of staff as Bill. But in the White House, you never heard the words *Barack* or *Joe* under any circumstances, not even from people who considered them close friends.

When it comes to talking about famous people, Hollywood is D.C.'s mirror image. I learned this when I was told that Scarlett and Kerry were still on board, but that Natalie had dropped out. Erik Smith, the strategist who assigned me to the actresses, was the one to deliver the news. He also told me the whole thing had been Harvey's idea.

I didn't need to ask who "Harvey" was. Harvey Weinstein, the legendary producer behind everything from *Pulp Fiction* to *Spy Kids 3,* was a major Democratic donor. A week earlier, he had decided Clint Eastwood's ramble deserved a response from Hollywood. Phone calls were made. Favors were called in. In no time, three actresses had been tapped to fly to Charlotte and address important issues head-on.

The campaign staff was more than happy to have well-known stars on the program. When it came to the speech itself, however, they devised a bait and switch. Rather than have them speak about issues, we would team them up to promote online watch parties for President Obama's closing-night address. My first draft of a script, which I'd sent over on Tuesday, was designed to do exactly that.

But now Natalie was out. This meant we'd have to tinker with the dialogue, and on Wednesday morning I joined a conference call with Kerry to regroup. I recognized her voice immediately from *Scandal*. But there was another voice as well, with a nasal gruffness that suggested its owner had begun chewing a steak in 1997 and was not yet through with the gristle. Harvey was on the line.

And he was pissed. After reading my draft about watch parties, Harvey had concluded that I was trying to manipulate him. This simply wasn't true. I was trying to help *other* people manipulate him. But I suspected the distinction might not make much difference. Each time Kerry offered a thought, the producer leapt in.

"Don't let him do this to you, Kerry! Don't let them push you around!"

My day did not improve from there. Returning to the locker room, I learned that Scarlett, too, had spoken with Harvey. Rather than work with me, she would now write her draft herself. I couldn't help but wonder what sorts of things were being said about me.

Then, just a few hours later, I found out. I was in a coffee shop, stress-eating a salted caramel brownie and downing a double espresso, when my phone began to vibrate. It was a 917 number I hadn't seen before.

"Harvey Weinstein wants to speak with you," said an assistant.

"Really? Are you sure?"

But there was no escaping it. There was a brief silence, and then the assistant's cheerful singsong was replaced by a now-familiar bark. "David Litt?"

Harvey began our conversation by informing me, in no uncertain terms, that I was an idiot who ruined everything. I was used to the Obamaworld style of criticism (not mad, just disappointed), and Harvey's high-volume fury rattled me. What if he was right?

But then Harvey told me I was in way over my head, and suddenly my spirits lifted. This second explanation made way more sense. There was no good reason for our conversation to be taking place. I was powerless to either help or undermine. All I could do—all I had to do—was suffer verbal abuse. Seen in that light, being hectored by Harvey Weinstein was like getting punched in the face by Muhammad Ali. At one point, with impressively dramatic flair, he asked whether I thought he might know a thing or two about storytelling.

"Of course," I said, hoping flattery might work. "I'm a big fan of your movies."

"Oh yeah? Do you know how many times I've been nominated for an Oscar? Three hundred and four! More than any studio."

Harvey continued in this vein for several minutes. Then he came to an abrupt halt, like a pair of windup toy dentures at the end of their run. "So," he said calmly, "let's give them all the time they need, and forget about these watch parties. Okay?"

My year in government had not made me any better at replying to powerful people. I tended to stammer, unsure how confident to sound. But to my surprise, Harvey's anger had been liberating.

"I think you're looking for someone who can actually make a decision," I said. "Why don't I find an important person and have them call you?"

I made good on my promise, and eventually it was agreed that all talk of house parties would be scrapped. The two actresses would still be limited to seven minutes, but they would now be free to give separate speeches, each in a three-and-a-half-minute block. For the first time all day, I relaxed. It seemed the worst was over.

It was not. It turned out I hadn't been the only one to receive a call from Harvey Weinstein. He had spoken one-on-one with Kerry, too. Like Scarlett, she was now spooked, unsure whether to trust me with her draft. Then, just a few hours later, Jeff got a Harvey call of his own. Before my boss could start speaking, the windup dentures were unleashed.

"You like speeches? Good, because you're about to hear one!"

You had to give Harvey credit for his opening lines. And for message discipline: among other things, he asked Jeff how many times he'd been nominated for an Oscar.

As the Wednesday-night program started, I still had a half dozen speeches on my plate. Secretary Duncan. Cristina Saralegui. DWS. But the Harvey/Scarlett/Kerry drama had completely sapped my energy. These seven minutes of remarks were occupying nearly 100 percent of my time. There was a brief respite around 10 P.M., when our team squeezed onto the floor to see Bill Clinton. But even this didn't last. Only minutes after I returned to the referees' locker room, Scarlett's draft landed in my inbox. It was heartfelt and well written. Unfortunately, it was also long—over one thousand words when it was supposed to be under five hundred.

In any other setting, no one would be annoyed to get four extra minutes of Scarlett Johansson's time. But her speech was the same night as the president's. We simply could not allow speakers to go long. While the rest of our team left the arena, desperate to catch some sleep, I stayed behind to trim the draft. By 2 A.M., I had a version I thought would make everybody happy. I sent it to Scarlett, left the arena, and had just reached the hotel lobby when I felt an unmistakable, sickening buzz. I checked my phone.

Edits!

vibrate

Sorry, no internet!

vibrate

Throughout the history of this great country
we have fought & stru

vibrate

ggled

It was Kerry, stranded somewhere without e-mail, sending me her draft in a series of texts. I waited for the deluge to finish—forty-five messages in all. Then, too tired to ask how badly she had exceeded the word limit, I plunged into a far from restful sleep.

TO MY SURPRISE, THE CONVENTION'S FINAL MORNING BEGAN WITH relief: Kerry's draft was on message! It was also approximately the appropriate length. But the good news ended there. When I e-mailed Scarlett to ask if she had received my revisions, I didn't hear back. I tried again after an hour. Same result. The actresses took off from Los Angeles. They landed in Charlotte. They were on the way to the arena. Still, no word. It was only a few minutes before speech prep that I finally got an e-mail, and even this wasn't from Scarlett herself. It was from a publicist.

Scarlett has decided to use her version.
Harvey says it will be fine.

I had no idea how to handle the situation. Luckily, Erik Smith did. He decided we would play good cop, bad cop. He also decided I would be the bad cop.

There was only one minor hiccup: he neglected to share either of these details with me. The moment both Scarlett and Kerry were in the room, Erik became a fountain of obsequious charm. We were *so* happy to have them. They were *so* important to the campaign. Then he turned to me.

"This is David. He'll talk with you about time."

Before Scarlett and Kerry could argue, and before I could protest, Erik ushered the three of us into a tiny cubicle down the hall. Giant bold lettering identified the office as belonging to our campaign manager:

JIM MESSINA

But this was entirely for show. Half the size of my DNC workspace in Washington, the cubicle was practically unused. The only things in the room, apart from two stunning celebrities and a twenty-five-year-old sweating through his dress shirt, were a printer and computer on a desk. I sat on one side. Scarlett and Kerry took the other. I knew I had only a moment to earn their trust.

"Let's start by printing out your drafts," I said. I sounded firm, yet pleasant. So far, so good.

Then I hit control-P, and my reassuring smile flattened. Hardware is not my strong suit. I am the kind of person who readily concedes that the Amish have a point. And now the screen was displaying a printer error, a porridge of letters and numbers I had never seen before. Summoning tech support was out of the question. I didn't know their number, and even if I did know it, I couldn't afford to lose the appearance of control. Instead, I made firm eye contact with the star of ABC's new hit drama and the woman Vulture.com named "The Smart Sex Symbol" of 2012.

"So," I said, "do either of you know how to fix the printer?"

I could tell immediately that we were no longer on a first-name basis. Scarlett Johansson and Kerry Washington looked at me like I had just asked them to brush my teeth.

I tried my best to recover, suggesting that perhaps printing out drafts wasn't necessary after all. But it was too late. While the actresses reluctantly joined me at the computer, any semblance of authority had vanished. After fifteen fruitless minutes, Erik poked his head in. He seemed pleasantly surprised to find me alive, but disappointed to find me unproductive. A minute after that, Jeff appeared, took Scarlett Johansson to a different office, and left me and Kerry Washington alone to figure out her remarks.

At first I wondered if this, too, was part of Harvey's strategy. *Strangle him with an extension cord, Kerry! It'll look like a printer accident!* But for some reason, one-on-one, everything snapped into place. Kerry's remarks didn't actually need much polish. In the patriotic atmosphere of the convention hall, it was far easier to explain how seriously we took limits on time. Just twenty minutes later, Kerry and I exited the cubicle with a draft. Just ten minutes after that, Jeff and Scarlett wrapped up her draft, too.

On the convention's final evening, I watched with my fellow speechwriters as both actresses delivered their remarks to loud applause. A few hours later, Barack Obama formally accepted the nomination for president of the United States.

He began by admitting that politics had lost some of its luster since 2008. "Trivial things become big distractions. Serious issues become sound bites. The truth gets buried under an avalanche of money and advertising."

But the rest of his speech urged us to remember that for all their frustrations, campaigns still mattered. Doing big things was exhausting. The opposition to change was dispiriting. Still, progress depended on our willingness to try.

"Our path is harder," President Obama told us. "But it leads to a better place."

He was referring to jobs, education, and health care—not tirades about Oscar nominations delivered over the phone. Still, I couldn't help but think about the past two days. So much of my energy had gone toward Harvey. Time that might have been spent on issues that truly mattered was spent e-mailing publicists instead. Where did that place me on the scale of moral flabitude? Was putting all that effort into pleasing one powerful person really the right thing to do?

But what if I looked at it a different way? Were a few hours of my time worth placating someone who paid the salaries of organizers in Florida, Pennsylvania, and Ohio? What if those organizers swung an election? What if that election got Zoe Lihn her heart surgery? What if that surgery saved her life?

This, I had begun to realize, was politics. Sometimes the answer depends entirely on the questions being asked.

8

THAT FIRST REAL TASTE OF BLOOD

Along with *gadzooks* and *hubba hubba,* the word *zinger* fell from favor decades ago. Yet like a species of stick insect once thought to be extinct, it can still be found thriving on a small rocky island in the English language. That island is presidential debate prep.

The moment I returned to Washington from Charlotte, the zinger machine went into overdrive. POTUS and Romney were set to face off for the first time on October 3. That gave us less than a month to prepare. Each day, I spent hours thinking of clever jabs or vicious comebacks. Then I'd send the best of the bunch to Favs, who was curating submissions from writers across the country.

I wish I could say the goal of these one-liners was to win an argument. It wasn't. The goal was to make POTUS look good on TV. For better or worse (mostly for worse), debates have become a reality show. Al Gore lost one by sighing on camera. Richard

Nixon lost an even bigger one by sweating where JFK stayed dry. I can't imagine most Americans would feel unsafe knowing the commander in chief occasionally sweats or sighs. Yet during the biggest nights of election season, these are the standards by which we judge. Imagine trying to explain this to a Martian. Imagine trying to convince it that this—a dog show, but for people—is what makes us the greatest democracy on earth.

And yet, even in a made-for-TV political era, the truth has a way of coming out. On September 17, my twenty-sixth birthday, *Mother Jones* magazine released a leaked cell phone video of Mitt Romney. I couldn't have asked for a better present. The footage wasn't quite in focus, but over the heads of well-heeled donors, you could spot the unmistakable jawline of the Republican nominee.

"There are forty-seven percent of the people who will vote for the president no matter what," he said matter-of-factly. "There are forty-seven percent who are dependent upon government, who believe that they are victims, who believe that government has a responsibility to care for them." It was appalling: a presidential candidate writing off nearly half the people he hoped to serve.

To those of us who spent time reading right-wing media, however, it wasn't a surprise. Under Ronald Reagan, conservative theory held that handouts for the rich were good for everyone. Wealthy people would use their tax cuts to buy expensive stuff. Middle-class people would get jobs making expensive stuff. Poor people would get jobs selling middle-class people cheap versions of the expensive stuff they made.

But in the age of Obama, something had seeped from the comments section of Fox Nation into the mainstream GOP. According to the new theory, the entire lower half of the income scale was full of leeches. These deadbeats were getting more than they deserved. Once a matter of economics, cutting rich people's taxes had become a matter of morals instead.

I don't know if Romney truly believed what he was caught saying. But it hardly mattered. When you lead a political party, you can't help but reflect its members' views. And Americans were horrified to see those views on camera. A week after the video, the Gallup tracking poll had us up by six points, one of our largest margins of the campaign. Our odds on FiveThirty-Eight, Nate Silver's website, rocketed above 80 percent. Rumors spread that our opponent's biggest donors were about to abandon him. Romney was toast.

And Democrats were celebrating. In the weeks after the 47 percent video, even the DNC cubicles seemed cheery. For the first time since July, I abandoned my French bakery and joined Devlin at work. Sure, a few strange reports were coming out of debate prep in Las Vegas. Why, for example, did POTUS feel the need to abandon his practice sessions and tour the Hoover Dam? But these were random oddities. No use letting them darken our mood. Besides, as debate night drew closer, I was told that one of my zingers was a consensus favorite among the senior staff.

"Why is Governor Romney keeping all his plans secret? Is it because they're just too good?"

Out of context, this line may not seem like much. But on debate-prep island, it was a rare specimen: snappy, on message, with just enough snark. I couldn't wait to see POTUS unleash it. It would demolish Romney on the spot.

On the night of October 3, I logged into a livestream from my DNC desk. My coworkers had long ago left the second floor for watch parties. Fluorescent lights flickered ominously in the abandoned halls. But I had never felt so confident. Romney was on the ropes. We didn't even need a decisive win on the debate stage—with a tie, or even a not-too-catastrophic loss, a second term would be guaranteed. And we had the best political performer in a generation on our side.

At 9 P.M., the candidates took the stage at dueling podiums—Romney in a red tie, President Obama in a blue one. POTUS spoke first. By coincidence, the debate fell on his twentieth wedding anniversary, and he began by addressing the First Lady in the crowd. "I just want to wish, sweetie, you a happy anniversary, and let you know that a year from now we will not be celebrating it in front of forty million people."

Cute, right? Not to me. Like anyone who had spent lots of time listening to President Obama, I knew immediately that something was off. The instrument was out of tune. That stilted, forced delivery? The word *sweetie* dropped haphazardly into the middle of a sentence? It was as though this were candidate school and Barack Obama was taking his oral exams.

Then our opponent spoke. "Congratulations to you, Mr. President, on your anniversary." On my computer screen I could see Mitt Romney's mouth moving, but surely the voice belonged to someone else. The candidate addressing POTUS was warm. He was charming. He sounded like a human. "I'm sure this was the most romantic place you could imagine, here with me." Where the audience had chuckled politely at President Obama's opening line, Romney earned an appreciative, genuine laugh.

My stomach turned.

For the next few minutes I forced myself to stare at my monitor, hoping President Obama would bounce back. Then, revising my strategy, I began pacing the empty hallway and muttering *fuck-fuck-fuck-fuck-fuck-fuck-fuck*. Romney was fighting for his political life, landing blow after blow.

POTUS, to put it mildly, was not. After almost two years spent watching President Obama, I knew when he was phoning it in. The annoyed half smile. The words *Now look,* at the beginning of every sentence. The halting "uh" sounds that could suck the soul out of a phrase.

Nor was it hard to guess why this was happening. For all his talent in front of the camera, POTUS hated political theater. At times, he reaped the benefits from refusing to play the game. But now he was paying dearly. With every lackadaisical answer, he breathed new life into Romney's campaign.

More than an hour into the debate, POTUS used my zinger. It failed to zing. Not long after, I turned off the livestream in a daze. I plodded down the second-floor hall, up the concrete steps, back through the third floor, and down the elevator. Then I drove to Jacqui's house. We spent the night cycling through unwelcome emotions: sadness, anger, confusion, despair. It was as though a friend had died testing a bungee cord he'd designed himself.

ROMNEY SHARP AND STEADY IN FIRST PRESIDENTIAL DEBATE

PRESIDENT OBAMA SNOOZES AND LOSES

OBAMA FLUNKED THE FIRST DEBATE. BUT WHY?

The headlines the next morning were brutal. But they weren't unfair. Trudging to my desk on the morning of October 4, I thought of everyone who had invested themselves in the idea of Barack Obama. My vols in Ohio. My coworkers in the White House. My fellow speechwriters from the convention. Scarlett and Kerry. Stacey Lihn. And at our campaign's most critical moment, just when we had turned a long-shot reelection bid into a sure thing, our candidate had acted like he didn't even care. This wasn't just a missed opportunity, or a subpar performance. This was an insult.

I was angry. More than that, I was heartbroken. The myth of Obama was not that he was somehow more than human. It was that he was the best possible version of a human, and that by following him—by believing without thinking, by rarely asking questions and never admitting doubts—I could become the best possible version of a human, too. Now I saw how stupid that was. POTUS was brilliant. He was talented. He was on the right side of history. But at the end of the day, Barack Obama was just a guy.

I still liked the president and believed he deserved reelection. But slumped in my cubicle, staring into Devlin's frightened face, I knew my relationship with my candidate had changed forever. My days in Obamaworld weren't finished. But my days as an Obamabot were done.

"WE CAN'T AFFORD TO LOSE."

As catchy phrases go, "Yes we can" was more exciting. But in the days after the debate, our old slogan failed to inspire. I had been drawn to politics in a kind of patriotic ecstasy. But what kept me there now was a list of pragmatic considerations, a bloodless preference for a concrete set of policy goals. Obama's reelection would insure more people. Create more jobs. Help more students pay off loans. It was the political equivalent of a vegan cookie—all nutrition, no taste. Still, it kept me from quitting my job.

Also, Mitt Romney scared me now. He still seemed like a fundamentally decent person. But the 47 percent video suggested that, as president, he would be governed by his party's darkest impulses.

This was not just a matter of rhetoric. Presidents in movies choose between good and evil, but real life rarely works that way. More often, presidents choose between easy and hard, between those who have power and those who do not. I thought of two-year-old Zoe Lihn, unknowingly waging a battle with her giant insurance company and the army of lobbyists it employed. I had no doubt whose side President Romney would take.

As the idea of losing became more frightening, it also became more real. On Nate Silver's site, our chances of winning had been climbing by 7 percent per month. After the debate, they fell 7 percent in just two days. I had no clue how long the slide would last. I refreshed the site obsessively, as if my clicks held the power to reverse the trend. They didn't, of course. By October 17, we had gone from a slam dunk to a virtual coin toss.

The only thing sinking faster than our poll numbers was morale. The entire DNC was summoned to an emergency meeting, during which we were told to stop refreshing FiveThirty-Eight. Then the pep talks began. "When the game's on the line," said one high-ranking aide, "the president's the guy you want holding the ball." A week earlier this might have reassured me, but not anymore. Had this guy spent debate night in a coma?

Other Obama advisors addressed the disaster as well. On a conference call, David Axelrod tried valiantly to shoulder the blame. He had set the strategy, he told us. He should have let POTUS be more aggressive in his attacks. I admired Axe's loyalty, but for the first time I found one of his arguments entirely unconvincing. *Of course he would say that,* I thought. *He's an Iowan.*

By this I did not mean David Axelrod was born in Des Moines or Cedar Rapids. I meant he was one of a few dozen people who were with Obama from the very start. For nearly a year the Iowans crisscrossed the Hawkeye State, down twenty points in the polls. It was the political equivalent of living in caves, surviving on rainwater and grubs. Cold nights. Long odds. The Butter Cow at the state fair. These were the stories they told when times got tough, the way George Washington's aides must have once regaled new conscripts with tales of Valley Forge.

I envied the Iowans' unflappability. Surviving a near-death

experience had put ice water in their veins. But left unchecked, their confidence could become delusion. I had seen it just two years earlier, when Democrats trailed badly in midterm polls. As conventional wisdom declared we were done for, Obamaworld veterans were summoned to lift our hopes. "Don't forget, this is exactly what they told us before Iowa," they promised. I understood their point. Sometimes pundits write you off because they can't imagine anything beyond the status quo. But other times pundits write you off because you're about to get your ass kicked. No one knew which kind of year 2012 would be.

This much was clear: for all his chest-thumping swagger in the Oval, POTUS had been wrong. Romney *could* beat him. And strangely, rather than worrying the president, this seemed to animate him. He was like a fighter who only gets going after that first taste of blood. Romney had landed a punishing blow. POTUS was through playing around.

He also seemed to reconsider his disdain for political theater. During the debate, Romney had proposed cutting funding for PBS. Now, President Obama pounced. "Elmo, you better make a run for it!" he cried in speeches. It was a silly line—something designed for TV cameras. But the president's passion was genuine, and crowds responded. So did I. For all my disappointment with his first-debate performance, I had no doubt President Obama understood how much depended on this campaign.

I also knew that, of the seven billion people on earth, no one hated losing more than POTUS. His competitiveness was legendary, even among his Type A staff. On one occasion, Hope Hall goaded the president into doing an on-camera impression of Frank Underwood, Kevin Spacey's character from *House of Cards*. When I asked her how she coerced the commander in chief, she smiled.

"Easy. I just told him I didn't think he could do it."

President Obama's deep-seated hatred of second place served

him well at the next debate. He came out swinging and, by a small yet decisive margin, was declared winner of the night. But then, just as we seemed to be regaining momentum, both sides hit pause.

It was time for the Al Smith Dinner.

ON AN OCTOBER NIGHT IN 1960, JOHN F. KENNEDY AND RICHARD Nixon ditched the campaign trail, donned white tie, and drove to the Waldorf Astoria hotel. There, they spent an evening telling jokes to wealthy New Yorkers. It's a tradition that has repeated itself nearly every four years since.

The Al Smith Dinner raises millions for local charities, but a noble cause does nothing to diminish the evening's strangeness. Imagine if, with five minutes left in the Super Bowl, the opposing quarterbacks rushed to the fifty-yard line to sing "I Will Always Love You" in a karaoke duet. Now imagine if, in addition to playing for opposing teams, each quarterback loathed everything the other stood for. Welcome to the Al Smith.

While Favs and Cody drafted more important speeches, I worked with our usual team of joke writers to assemble a draft. After months of intense campaigning, we had plenty to make fun of. Our opponent's stiffness. His gaffes. The 47 percent video. My script referenced them all. I even asked our advance team to create a "binder full of women," a prop inspired by one of Romney's second-debate gaffes. But on the day of the speech, when Favs took the draft to the president, POTUS cut nearly every aggressive joke. There would be no binder on the podium that night. My mind instantly went to the days before the first debate. Was he pulling punches again?

But this time, I was wrong to second-guess the president's instincts. As I watched the competing monologues from the comfort of Jacqui's couch, my attention turned from the speakers to the audience. The room was a sea of strained smiles. A complicated code of etiquette writhed just below the surface. For the first time, I realized how risky my initial draft had been. To throw bombs would have been satisfying. But it also would have made the president seem desperate and cruel. POTUS was playing the long game. Where I had tried to win an evening, the president was trying to win a campaign.

So was Romney—and to my great disappointment, he was doing okay. Like the Iowans, he had come face-to-face with political mortality, and it stripped away his fear. "We're down to the final months of the president's term," he said at the Al Smith Dinner, face beaming with a confident grin. The crowd, mostly white and wealthy, applauded with relish. A few people even whistled.

"He looked like a president," wrote *Wall Street Journal* columnist Peggy Noonan not long after. "He looked like someone who'd just seen good internals."

Perhaps he had. But while Peggy Noonan may not have known it, Obama had seen good internals, too. Where public polling found the race essentially even, Joel Benenson's numbers showed us reclaiming our narrow lead. Even more important, with just a few weeks until Election Day, the pool of undecided voters was shrinking. The campaign wasn't over. But the time for persuasion had passed.

In consequence, the number of remaining speeches began to dwindle. Favs and Cody would remain busy through Election Day, but as rhetorical handyman, my work was drying up. I wasn't the only one. At DNC headquarters, cubicle after cubicle was abandoned as employees were sent into the field. In typical, militaristic campaign fashion, this is known as "being deployed." With just weeks remaining until Election Day, and not much writing to do, I began to plan for a deployment of my own.

In late October, I completed my final speechwriting assignment of the 2012 election: a Q&A for *Us Weekly*. In "President

Barack Obama: 25 Things You Didn't Know About Me," the commander in chief informed America that he was left-handed, that apples were his favorite healthy snack, and that he enjoyed both bodysurfing and shaved ice. If we lost, this would be my farewell address. Not long after sending in my draft to Favs and Cody, I dusted off my giant green backpack from college. Then I hit the road for Cleveland, my home for the campaign's final days.

THE CAMPAIGN ASSIGNED ME TO A ROOM IN A VOLUNTEER'S HOUSE, a two-story white Cape Cod, and I arrived sometime shortly before midnight. The moment I rang the doorbell, I heard a deep, murderous growl from within. There was a crash of giant paws. Then the softer padding of slippers. My cheerful host, Norma, pried open the door.

I noticed she was middle-aged, with hair in curlers and a terry cloth robe. But the bulk of my attention was devoted to the beast behind her. A cross between a Great Dane and Satan, its giant blocky head raged with drool. I don't remember the dog's name. I do, however, remember the conversation between my host and me.

NORMA: Don't worry. Monster loves people. Come on in!
MONSTER *(translated):* I'LL RIP YOUR ARMS OFF, YOU LOUSY BASTARD!
NORMA *(laughing):* Really? You're scared of my puppy? I suppose I can put him in the bedroom.
DAVID: Yes, please.

Confinement didn't suit Monster, who immediately began beating his paws against the door in an escape attempt. Before he could succeed, I rushed my backpack up the stairs and into my room on the second floor.

The following morning, I sprinted in terror from my bedroom to the car. Then I drove to an Obama for Ohio office on the city's east side.

In campaign euphemism, I had been assigned to a "base area." In English, that meant residents were overwhelmingly African American and that many of them didn't vote. Every four years, well-meaning white people appear in these neighborhoods like cicadas. They knock on doors. They hand out flyers. Then they return to their burrows to eat acai berries and purchase Jonathan Franzen novels they'll never read.

The Obama campaign hoped to replace this unseemly practice with something better: self-sustaining, locally run organizations, links in a nationwide chain. In some places, the plan worked. Where it didn't, people like me were called in to "layer" the organizers in charge.

I wish I could tell you I recruited all my volunteers locally. But with numbers to hit, and little more than a week until polls opened, there was no time for the high-minded principle of 2008. A few days after I arrived, Jacqui arrived with two friends in her ancient fiberglass Saturn. Others soon joined our yuppie militia, bidding farewell to their Brooklyn nonprofits or taking leave from well-paid jobs. Even the supporter housing was bourgeois. Jacqui was assigned a room in a mansion belonging to a pair of cheerful white attorneys and their lovable Brittany spaniel.

Yet this style of campaigning, while hardly poetry, had its moments. One morning my friend Stephanie was sent to knock on doors with an eighty-four-year-old woman named Beulah. As they left the office, the redheaded Brooklyn lesbian in vintage clothing towering over the shrunken matriarch known locally as "Mother Carter." I wondered if we weren't asking too much. But the next day Mother Carter returned in her orthotic walking shoes. She would knock on doors all afternoon, she said—just as long as she got to team up with Stephanie.

That's the best part of fieldwork: with so much on the line, bridges build themselves. And as they do, the entire universe shrinks to just a few square blocks. For the first time in months, it was easy to ignore Nate Silver. In fact, I didn't check FiveThirtyEight until the final Friday of the campaign. To my relief, I learned our odds were improving. Victory was once again within our reach. More at ease than I'd been in weeks, I fell into a peaceful sleep.

I awoke at 2 A.M. Something wasn't right.

In my sleepy haze, it took a moment to figure out the problem. It was the outside of my stomach: my skin felt like it was on fire. Just the aftermath of some strange dream, I figured. Still, to be safe, I turned on the lights and lifted my shirt.

I had never seen anything like it. Across my abdomen were row after row of red bumps. In the center of each was a puncture wound. It looked like I'd been assaulted by a madman with a thumbtack. With horror, I recalled Norma telling me the dog often slept in this bedroom. In a worst-case scenario, I had bedbugs. In a best-case scenario, I had fleas.

Pivotal moments in a relationship are rarely the ones you expect. On our first date, I wondered what Jacqui looked like naked, and whether we had the same taste in TV. Never once did I ask myself, "Is this someone I can turn to in case of infestation?" Now, one year later, I could think of nothing else. When she answered my 2 A.M. phone call, I practically wept.

"This is going to be fine," she told me, although I could tell she didn't believe it. "Here's what we'll do."

If you're like me, you have never disposed of a body. After that long night in Cleveland, however, I'm confident I could. From a cabinet in Norma's bathroom, I pilfered trash bags and duct tape. I put my clothes in the bags, tied them tight, and covered the knots with tape. The trash bags went into my backpack. My backpack went into more trash bags. More knots were tied. More tape was taped. Once outside, I placed everything in my trunk and sped off.

Jacqui was waiting for me at the doorway of her mansion, and she immediately took charge. On her orders, I stripped to my underwear and ran straight for the shower. Then I stuffed as much clothing as I could in the dryer, leaving the rest in the car. Once our DIY extermination was complete, I passed out on the deliciously soft sofa in the living room, dreaming of lies to tell my new host and hostess in the morning.

With that, Get-Out-the-Vote weekend began. The final four days of a campaign are always a blur, and these were even blurrier than usual. There was the persistent unease that came with parachuting comfortable white people into a neighborhood they were unlikely to revisit. There was the realization that, despite these moral compromises, we still did not have enough volunteers. And there were the insect bites, which stung rather than itched and grew more painful every day.

With so many sources of discomfort, we had no time to reflect on what was happening. We knocked on every door in the neighborhood once during the weekend, once more on Monday, and twice on Election Day itself. Before we knew it, polls were just half an hour from closing. That's when my BlackBerry vibrated.

Run!

It was the e-mail sent to all field staff in the waning moments of a campaign. Grabbing walk packets, Jacqui and I drove to a row of small brown houses, where we each took a different side of the street. I knocked on door after door, searching for would-be voters. No luck.

Finally, at 7:26, just four minutes before polls closed, I heard shuffling from a living room. A woman appeared in a robe and sandals. She was registered but hadn't voted yet. I was thrilled.

"You've got to go now!" I told her.

"Nah-ah," she said. "I'm not voting."

"But the election's going to be very close," I reminded her.

"I told you," she said. "I don't vote."

This was truly a fairy tale ending. A nonvoter! A last-minute conversion! All she needed was someone to believe in her.

"This election is *so* important," I implored. My voice dripped with emotion. "Ohio's going to be *so* close. Your vote could make a *big* difference."

In the movie version of my life, this was the moment when she raced to her polling place. In the my-life version of my life, it was the moment when she called me an asshole and told me to get off her porch. And with that inspiring exchange, time was up. The reelect was done.

THE CUYAHOGA COUNTY DEMOCRATS HELD THEIR WATCH PARTY AT a DoubleTree in downtown Cleveland, and Jacqui and I arrived just as the first few states were being called. The first few were unsurprising: Vermont blue, Kentucky red, and so on.

Soon, however, we allowed ourselves to hope. Early in the night, CNN projected Obama would win Pennsylvania, a state where Romney had made a last-minute push. Then we won New Hampshire, another potential tipping point. North Carolina went to our opponents, but we took Wisconsin, exactly as Joel Benenson's internals predicted. The nervous chatter in the DoubleTree was replaced with a prayerful silence. If Ohio slid into Obama's column, it would all be over.

Ever since the first debate, I had stubbornly refused to think about winning. But now, eyes glued to a TV screen in anticipation, I granted myself the tiniest moment to reflect. Unemployment was near 8 percent. Obamacare wasn't popular. Despite his once-in-a-generation talent, our candidate had belly-flopped on his campaign's biggest night. And yet, despite everything, we stood on the brink of a second term. What better vindication of our efforts could there be?

Ding-dong-thunk-thunk-THUNK-THUNK-ding-dong.

A series of metallic sounds, half chiming, half clanging, echoed through the ballroom. Wolf Blitzer had a projection to make. "CNN projects that Barack Obama . . ." he said. But no one waited for him to finish. Beneath a picture of POTUS were the words I had scarcely allowed myself to imagine.

REELECTED PRESIDENT

The room exploded. It wasn't a cheer. It wasn't a scream. It was a supernova of relief. I threw my hands in the air. Jacqui turned toward me, laughing uncontrollably, and wrapped me in a hug. The noise faded just enough to hear Wolf Blitzer continue.

". . . because we project he will carry the state of Ohio."

The ballroom erupted all over again. "Yes we can! Yes we can!"

And then, spontaneously, our time-honored chant evolved into something I had never heard before. "Yes we did! Yes we did!"

I was almost too tired to cry. I was almost too happy to speak.

Only now, with victory in hand, did I realize that part of me thought this day would never come. Anything can happen once, even electing a progressive black man named Barack Hussein Obama president of the United States. But twice? Like so much I had seen over the last four years, it was impossible, and then it was still impossible, and then it had been done. Now, thanks to a remarkable victory, we had four more years to change America. Four more years to close the gap between the country we were and the country we hoped to be.

Was it 2008 all over again? Of course not. I had learned too much about delay and disappointment since then. Running my hand across my punctured belly, I was reminded that some scars don't quickly heal. But jumping for joy with Jacqui, thinking of the years to come, I wished I could summon Sarah Palin to my side.

"The honeymoon phase may be over," I'd tell her. "But that hopey, changey thing? It's working out just fine."

PART TWO

OUR (TEENSY) PLACE IN HISTORY

9

HITLER AND LIPS

On June 4, 2005, a skinny young senator delivered the commencement address at Knox College, a small liberal-arts school in Illinois. In Obamaworld that speech is like *Star Trek: Deep Space Nine*. Casual fans may not have seen it, but it's worshipped by the hard core.

Rewatch the Knox commencement today, and here's the first thing you'll notice: the students are remarkably calm. Sure, there's some cheering and clapping, but no one's freaking out about Obama. A few audience members are doubtless wondering who he is. And with his opening anecdote, the speaker acknowledges his newness on the national stage.

"I have not taken a single vote, I have not introduced one bill, had not even sat down in my desk, and this very earnest reporter raises his hand and says, 'Senator Obama, what is your place in history?'"

Here the young lawmaker pauses, a hint of the timing that will later serve him well.

"I did what you just did, which is laugh out loud."

But if the senator's amusement is genuine, so, too, is the sincerity that follows. In fact, after reflecting on the reporter's question, he surprises the students with the central theme of his speech.

"What," he asks them, "will be your place in history?"

This was part of what first drew me to Obama: he turned grandiosity on its head. For as he told the Knox College graduates, throughout most of human history, your destiny was certain. Your fate was sealed the moment you were born. America changed that. What made us special—what made us exceptional—was the promise that ordinary people could shape the national life. In fact, they were expected to. For 229 years it was our audiences, not our speakers, who made our country great.

What will be your place in history? In America, that's not such a dumb question after all.

As President Obama's second term began, it was certainly the question on everyone's mind. Many exhausted top advisors, their personal legacies cemented and a second term secure, were finally ready to leave. David Plouffe, the electoral mastermind. Deputy Chief of Staff Nancy-Ann DeParle, who stewarded the president's health care law. Axe, who shaped the Obama vision more than anyone except Obama himself.

And Favs. At just thirty-one years old, Jon Favreau was already one of the most accomplished speechwriters in history. With no new rhetorical worlds to conquer, he left in early 2013.

I was sad to see Favs go. He was a good boss and a remarkable talent. But President Obama's loss was my gain—the chief wordsmith's departure left an opening at the bottom of the speechwriting totem pole. When I returned to the White House in March, my days of professional promiscuity were over. I was now a full-time member of the president's team.

At the top of the totem pole, Favs's spot was filled by his deputy, Cody Keenan. It was a smooth transition: Cody had been writing speeches for POTUS since the first campaign. Still, there were subtle differences between my old boss and my new one. I always thought of Favs as an architect, carefully crafting structures of great delicacy. With Cody, the phrase *warrior poet* sprang to mind. Like his mountain-man beard, which sprouted whenever the stakes grew high, his writing began with something visceral inside him. Emotions came first, only later to be trimmed. This difference in writing style was reflected in attitude. Favs's confidence was tinged with cleverness; Cody's with righteousness.

I was especially grateful for this righteous swagger during a POTUS meeting my first week back. The subject was the Gridiron Club Dinner, which, together with the Correspondents' Dinner and the Alfalfa, makes up the Holy Trinity of presidential comedy events. It had now been eighteen months since I first entered the Oval. I told myself I wasn't nervous. But when I reached the doorway I froze, terrified of doing something dumb.

Fortunately, Cody was completely comfortable. The president was seated behind his desk, and my new boss sauntered toward him like a detective about to crack a case. I tiptoed cautiously behind.

"So," POTUS asked, "are we funny?"

This was less a question than an invitation to make small talk. Cody didn't miss a beat.

"Well, Litt's pretty funny," he said, nodding in my direction.

A brief hint of confusion crossed the president's face. He clearly wasn't sure he'd heard right. But after a moment's pause, he decided to keep going.

"Yeah," POTUS said. "Lips *is* funny."

As you might imagine, I have replayed this moment frequently in my head. Perhaps I simply misheard the president. Perhaps time has warped my memory. But I don't think so. I'm fairly certain Barack Obama called me *Lips*.

Here's why I'm so sure. Number one, POTUS enjoyed banter. Going out of his way to extend pre-meeting small talk is exactly the sort of thing he would do. Second, while I wish I could say otherwise, President Obama had no reason to know my name. I'd written scripts for tapings. I'd helped him with a Correspondents' Dinner. But lots of people pop in and out of a president's orbit each year. At most I was a familiar face, the barista at your local Starbucks, the robber who wasn't Joe Pesci in *Home Alone*.

Finally, it was not the first time this sort of incident had occurred. One friend, a fellow comms staffer who started at the White House in 2009, was named Jason. POTUS loved his work and would often ask his boss to pass along a thank-you. "Tell James, 'great job,'" he would say. There was nothing anyone could do about this. After a few years, the president figured out Jason was Jason. Until then, James was James.

So Lips it was. And to be honest, as I settled onto the couch in the Oval, I wasn't embarrassed. I was thrilled. The president of the United States had referred to me by name! True, it wasn't *my* name, but no need to get nitpicky. Besides, I liked having an alter ego. Litt was shy. Litt was timid. But Lips? Lips could be bold. Lips could be daring.

Lips didn't give a fuck.

What better way to describe the growing sense of liberation I'd felt since the second term began? After the inaugural ball, for example, the wait at the coat check had stretched for ninety minutes. This wasn't surprising; the coat-check line at the first inauguration was similarly dysfunctional. But back then, I would never, under any circumstances, have cut in front of Obamaworld's elite. The building could have been on fire. The ballroom could have been swarming with killer bees. I still would have waited my turn.

Now, things had changed. Thanks to a well-connected friend, Jacqui and I had tickets to Rahm Emanuel's after-party at an underground blues club. And thanks to my spot on the Presidential Inaugural Committee, I had a plan. Ducking into a staff room, I grabbed a curly rubber earpiece, the kind used by Secret Service, and switched off the walkie-talkie. Then I held my hand to my mouth as though a microphone were clipped to my sleeve.

"Four-zero-niner, we have a garment situation in progress. Over."

An older lady in an expensive gown gave me a look suggesting I had recently crawled out of a sewer. Then she saw my earpiece, apologized, and immediately stepped aside.

"VIP coat retrieval, roger."

"Priority jacket acquisition, ten-four."

Five minutes later, I was at the front of the line. Five minutes after that, I returned, bearing my girlfriend's coat. And five minutes after that, we were zipping through the freezing night air in a pedicab, on our way to watch Chicago's mayor sway awkwardly as a lineup of living legends played the blues.

Classic Lips.

WHICH BRINGS ME BACK TO THAT OVAL OFFICE MEETING—MY FIFTH, Lips's first—during which the Gridiron Dinner was discussed. In 1885, when the Gridiron Club was founded, most of its members were grouchy old print journalists. Today, they still are. At their annual spring meeting, guests wear white tie. Reporters don costumes and perform parody songs about the politicians they cover. Petits fours are eaten. The evening ends with a chorus of "Auld Lang Syne." It's a four-hour homage to what people did before TV.

While nothing compares to journalists in Elvis wigs singing "Block Barack Around the Clock" (this actually happened, in 2011), the evening's real highlights are the guest speakers. By tradition there's a Republican, a Democrat, and, if the invitation is accepted, the president of the United States.

Like the rest of the program, these monologues are a throwback to a simpler time. Even the Gridiron's motto—Singe, but Never Burn—is an artifact of a happier, pre-Internet age. The draft I sent to POTUS included such scintillating topics as "Budget Cuts," "Press Conferences," and "Guys Named Gene." I can't say he laughed uproariously at any of the jokes. But he understood his audience. Pronouncing himself satisfied, he showed us to the door.

I never dreamed of grabbing an Oval Office apple. Even as Lips, my cockiness had limits. Still, as I left the meeting, there was brashness in my step. I had been back at the White House less than a week. Already, I was meeting with the president himself!

As if that weren't enough, the building was being restocked. Eager twenty-two-year-olds, fresh off their first campaign, were filling the entry-level jobs. Shyly, the newcomers would ask if I could get coffee sometime. When I agreed to, they practically melted with relief.

"Don't worry," I'd tell them. "Networking's only bullshit if you're bullshit."

To my surprise, however, I also had real pearls of wisdom to dispense. These had nothing to do with career ladders and everything to do with the unique environment in which we worked.

"Don't tell them I sent you, but the fifth-floor librarians keep a bowl of candy by the reference desk."

"Steer clear of the gym on Tuesdays. It's Zumba night."

"You know the medical unit in the basement? You can take all the free Advil and Band-Aids you can carry. They'll even give you Sudafed if you sign for it."

The wide-eyed youngsters would thank me. Then I'd return to my desk, the drawers stocked with candy and over-the-counter pharmaceuticals, to continue working on my Gridiron draft.

By the middle of the week the jokes were almost finished, and I turned to what we called "the serious close." These are staples of humor speeches, two or three paragraphs of sincerity at the end of otherwise lighthearted remarks. Since the audience was full of reporters, I took the occasion to praise journalists who embodied the best of the free press.

"They've risked everything to bring us stories from places like Syria and Kenya," I wrote, "stories that need to be told."

For a moment, I wondered if I should run this line by a foreign-policy expert. That's what Litt would have done. But then I thought better of it. Lips didn't need some egghead to tell him how to craft a sentence. You lump the countries ending in *yuh* sounds together. Everything flows perfectly. The crowd goes wild. The end.

On Saturday night, my confidence was rewarded. Dressed in a tailcoat, wing-collared shirt, and white bow tie, I descended the escalator to a ballroom in the Renaissance Hotel. Just feet from the stage, I watched costumed reporters sing "My Gun" to the tune of "My Girl." Louisiana governor Bobby Jindal and Minnesota senator Amy Klobuchar delivered monologues in front of a giant, glowing spatula I can only assume was the club's namesake Gridiron.

Then, just before "Auld Lang Syne," POTUS spoke. The laughter was consistent and chummy. The jokes singed, but never burned. During his serious close, President Obama praised reporters for risking everything in places like Syria and Kenya. Everything flowed perfectly. The crowd went wild.

Lips was fucking crushing it.

On Monday, back in the office, I was a paragon of false humility. Other speechwriters told me the remarks had been great, and I replied that it was all in the delivery. I explained to the twenty-two-year-olds that, really, when you think about it, credit belongs to the entire team. I was on top of the world, wondering if I might actually be the best wordsmith in history, when I heard from one of the president's longest-serving speechwriters, a soft-spoken Massachusetts native named Terry. He had a question.

This was a bad sign. At nearly forty, Terry was our team's elder statesman, and he had a talent for leading witnesses that only lawyers and parents of young children possess. If Terry wondered whether your smoke alarm was working, your house was on fire. If he asked how your girlfriend was doing, you could safely assume she was, at that very moment, enjoying sweaty intercourse with your best friend. Now Terry wanted to know if I'd seen an article in the *Daily Nation*, a newspaper published in Nairobi.

KENYA NOT SAFE FOR FOREIGN JOURNALISTS, SAYS OBAMA

It took some frantic googling, but I pieced together what had happened. The White House press office had released the full transcript of POTUS's Gridiron speech. When Kenyan officials read it, they noticed their country featured in the same sentence as one of the world's most despised regimes. They went berserk.

Ordinarily, that would have been the end of it. But because this was international diplomacy, the Kenyans used their newfound berserkness to gain leverage over the United States. Bitange Ndemo, Kenya's Permanent Secretary for Information and Communications, released an official statement calling the president's words "not only inaccurate, but exceedingly disturbing." A group called KOT (Kenyans on Twitter) created a brand-new hashtag to channel their rage.

If you've never angered a country of more than forty-five million people before, it might seem like a power trip. It's not. Sitting in my office, typing *Kenya at Obama* into the search bar over and over again, I felt more helpless than ever. What I wanted more than anything was someone I could talk to—a Kenyan I could call. "You think *your* entire country is mad at *my* entire country," I would explain, chuckling at the mix-up, "when really it's just little ol' me."

But no such Kenyan existed. As a junior White House staffer, I had the ability to place the salmon of distrust in the toilet of international relations. I didn't have the ability to get it out.

Instead, senior staff were forced to busy themselves playing cleanup. These were people with serious responsibilities, global priorities on their plate. Now they had to take time away from far more worthy objectives to deal with my mess. Even then, it took an official apology on America's behalf to put the controversy behind us. "We recognize and commend the press freedoms enshrined in Kenya's constitution," said an unnamed White House official. "Obviously, the situations in Syria and Kenya are quite different."

Perhaps Lips was not crushing it after all.

I spent the rest of the day wondering if antagonizing a midsize African nation was considered a fireable offense. But in the White House, I learned, it's not so cut-and-dried. I wasn't the first young staffer to find their smallest screw-up magnified to national scale. I wasn't even the first speechwriter to piss off another country by mistake.

Besides, Obamaworld wasn't big on firing people. Exceptions were made for the grossly negligent or the publicly embarrassing. But the merely incompetent, rather than being dismissed entirely, were simply exiled from a circle of trust. It didn't happen immediately. Instead, like a pirate marooned on a desert island, you found yourself watching as everything you ever cared about slowly disappeared. Eventually, your responsibilities vanished completely. A few lackluster employees were fine with this arrangement, cashing steady paychecks for an ever smaller amount of work. But most took the hint and resigned.

Living, as I was, in fear of being cast away, I desperately craved a shot at redemption. The days crawled by. No second chance arrived. Finally, at the end of March, Cody gave me an assignment. POTUS was scheduled to speak in Miami, where a new tunnel would soon connect the city and the port. The remarks were mine.

Not that anyone else wanted them. With the possible exceptions of toenail clippings and the Professional Bowling League, infrastructure finance is the least sexy subject known to man. Yet this only made me more eager. If I could write something riveting about the world's most boring topic, surely I would be in my team's good graces once again.

IN THE OBAMA WHITE HOUSE, THERE WAS NO SET TIME LINE FOR drafting a presidential speech. For an unexpected eulogy, the writer might have just forty-eight hours. The State of the Union took six weeks or more. But for my remarks in Miami, which were as typical as they come, I had about seven days.

My work began with a policy meeting. This is a polite way of saying that a team of experts tried, often unsuccessfully, to stuff information into my fluff-filled head. I can't say I always enjoyed being the token ignoramus. But there was power in the role. If a concept couldn't be explained to me, it was unlikely to make it into the speech.

The next step, after absorbing my colleagues' wisdom, was independent research. I would later learn to focus these efforts on the interesting, rather than the technically important. But for my Miami speech, desperate to restore my reputation, I embarked on an all-day infrastructure binge. Advances in tunnel-digging technology. Incentive structures associated with various investment schemes. Dredging. By the end of my self-taught crash course, I was hooked.

Holy cow! Public-private partnerships are awesome!

I'd later realize this was a warning sign, an early-onset symptom of lost perspective. But at the time, I was thrilled.

With research out of the way, I was finally ready to sit down and write. This was my first policy speech for POTUS since my return, but to my surprise, everything seemed easy. The previous forty-eight hours had left my brain completely saturated. Now, I poured that esoteric knowledge into a draft. I had written the race speech for infrastructure finance. I was sure of it. Two nights before the remarks were to be delivered, I sent them to Cody for his edits. Then I went on a Kennedy Center date with Jacqui and waited for my boss's applause.

In the EEOB, we had a term for what happened next, and it described both the condition of the speech and the ego of the speechwriter responsible. I got blown up. I had expected Cody's tracked changes to be minor flecks of red, the aftermath of a careless shave. Instead, I got the aftermath of a chain saw massacre.

Most painful of all, where ego was concerned, was that the new remarks were exponentially better than the old ones. Cody connected Americans' stories and America's story, self-interest and the national interest, in a way I simply had not. Reading through the wreckage, I wasn't mad. I was embarrassed. In my world, there was nothing worse than being useless. My Kenya fiasco, while embarrassing, could at least be written off as a mistake. But producing something worthless? Wasting my boss's time? In the Obama White House, these were character flaws. And they were unforgivable.

The remaining steps of the speechwriting process sped by in a demoralizing haze. Only Cody and I knew about the rewrite. My name was still on the remarks. But the privacy of the shame made it that much harder to bear. The day before the speech was delivered, policy teams, fact-checkers, and lawyers examined language they thought was mine. Their edits, by and large, were helpful. Their congratulations made me want to crawl into a hole.

I still had one more step remaining. That night I sent the remarks to "the book," the thick binder POTUS read each night, and in the morning I was informed he had only minor changes. The other speechwriters sounded impressed. "That's great!" they told me. I smiled weakly, wondering if they suspected the truth. Either way, the ordeal was almost over. Just a few hours later, President Obama was in Miami, reading what was still technically my speech.

"What are we waiting for?" he said. "There's work to be done; there are workers who are ready to do it."

It was classic warrior poet: blunt, passionate, steeped in common sense. My original draft had been far more concerned with cruise-ship disembarkation rates and high-tech hydraulic drills. It was only as I listened to the crowd cheer for Cody's lines that I finally realized the full extent of my mistake. I had written a draft of a speech about infrastructure finance. For a cabinet undersecretary or deputy mayor, that might have been fine. For the president, however, every speech is a speech about America. Every audience is the entire United States.

I had learned a valuable lesson. But had I learned it soon enough? A month after returning to the White House, the president's Knox College commencement had never seemed more relevant. "What will be your place in history?" I could hear him ask. From my blown-up draft, a million tiny whispers echoed in reply.

Desert island. Desert island. Desert island.

IF YOU'RE LOOKING FOR A NEAR-PERFECT OBAMA SPEECH, ONE THAT should be canon but isn't quite, I recommend the remarks he delivered on April 18, 2013. Three days earlier, a bomb had gone off at the finish line of the Boston Marathon. Two hundred sixty-four people were wounded. Three were killed. When POTUS arrived in Boston for the memorial service, the bombers had yet to be caught. His speech that day, written by Terry on impossibly short notice, had to be flawless, and it was. An extraordinary blend of toughness and tenderness, it's the kind of thing that earned POTUS the moniker "consoler in chief."

Yet as a nation mourned and a city searched for suspects, it was "comedian in chief" that occupied my time. The 2013 Correspondents' Dinner was just two weeks away.

It was a strange feeling, watching news about manhunts and shootouts while trying to write something snarky about Maureen Dowd. And along with the fear and discomfort felt by the entire nation, there was a question I couldn't shake. Since returning to the White House, I'd been given two major assignments. I'd bungled both. Would I once again screw things up?

It wasn't supposed to be this way. With Favs's departure, I was more responsible than ever for the Correspondents' Dinner, and should have been relishing my new role. Instead, in the two years since being hired by the White House, I had never felt so utterly fraudulent. I was living the exact opposite of a superhero's life. By day, I was a mild-mannered speechwriter. By night, I was a mess.

My inability to calm myself only added to my sense of phoniness. Unless I was forgetting something, *The West Wing* never did an episode where Rob Lowe buys a mouth guard because he grinds his teeth in his sleep. If it had, the writers might have learned what I did: the CVS near the White House sells two types. There's a no-frills mouth guard for about twenty dollars.

There's also a fancy, ergonomic model for forty. I didn't have to think twice about springing for the deluxe.

The old saying goes that you should fake it till you make it, but I think a more accurate saying would be "Fake it because you have no choice." Coming up with punch lines. Collecting jokes from our growing diaspora of writers. Convincing everybody that a high-concept short film—starring Steven Spielberg, Tracy Morgan, and Barack Obama playing Daniel Day-Lewis playing Barack Obama—was a good idea. Each day I went to work and did the best I could, not because I was courageous, but because I couldn't come up with an alternative. Then I went home and ground my high-tech mouth guard to a pulp.

Before I knew it, Cody was once again sauntering into the Oval, and I was once again tiptoeing behind. I felt good about our draft, but this no longer offered comfort. I had felt good about the Gridiron and Miami speeches, too.

The joke that most worried me involved the political landscape postcampaign. "One thing Republicans can all agree on after 2012 is that they need to do a better job reaching out to minorities," the script read. "Call me self-centered, but I can think of one minority they could start with."

It was the kind of line we never would have written in the first term. No one could remember POTUS referring to himself as a "minority" before. But with the reelect behind him, President Obama was eager to push the envelope.

"That's pretty good," he chuckled. Just as he did a year earlier, when the subject was eating pit bulls, he even promised a personal touch.

"I might add a little wave there. Maybe a 'hello,' or something." How strange. There I was, sick with nervousness, and POTUS was having fun.

While I doubt President Obama looked forward to spending

his Saturday night with the press corps, I always got the sense he enjoyed reading jokes. Unlike most politicians, President Obama missed being treated like a normal person. I once overheard him say that this was why he loved meeting babies: they had no idea who he was.

But everyone else did. Even five-year-olds could recognize Obama. Without quite intending to, POTUS had become the most famous person on earth. And he paid a price for it. Going for a walk. Eating at a restaurant. Catching a movie. Activities the rest of us take for granted were for Barack Obama a distant memory. It's not like joking with staff in the Oval was entirely casual. No one forgot which person was the president. But it was as close to normalcy as POTUS was likely to get.

That doesn't mean he laughed at everything. One of my fellow speechwriters, Kyle O'Connor, had written a joke about the cliquey nature of the Senate. It was a great idea, but required a Valley-girl accent to work: "So I found out that Mitch was talking with Rand who was talking with Lindsey who said that John and Ted were, like, talking?"

President Obama gamely gave it a try, but it was a disaster. He sounded forced and awkward, like a grandmother accenting the last syllable of *Beyoncé*. Throwing caution to the wind, I jumped in with my best Alicia-Silverstone-in-*Clueless* impression. POTUS's expression remained grim.

"That might be funny . . . if a comedian did it."

I swallowed hard, wishing I had stayed silent. We cut the joke from the script.

The rest of the meeting passed without incident. President Obama approved some jokes, told us to make others edgier, and sent us on our way. But surviving ten minutes in the Oval couldn't erase my feelings of fraudulence. The days of proudly strutting back to my office were over. The days of dark, harrowed circles under my eyes had begun. At work I continued faking it. What else could I do? But Jacqui, regularly awakened by my tossing and turning, knew the full extent of my fear.

A few days before the dinner, a second meeting gave us a chance to present the edgier material POTUS had requested. ("A book burning with Michele Bachmann. I like that!") It also gave us a chance to show him slides. Not everyone approved of displaying silly, photoshopped pictures in the middle of the president's monologue. Compared to well-delivered setups and punch lines it felt like cheating, and to some extent it was. But if years of writing have taught me anything, it's that people hate words and love pictures. Why not give them at least some of what they want?

Besides, POTUS liked the slides almost as much as the audience did. He laughed when he saw his face on a cover of *Senior Living Magazine*. He appreciated our putting a "Blame Bush Library" next to the real thing.

Most of all, he enjoyed a series of three pictures photoshopping the First Lady's new hairdo—eyebrow-length bangs—onto his head. POTUS with bangs, in front of an American flag. POTUS with bangs, relaxing alongside his wife. POTUS with bangs, walking side by side with Israeli prime minister Bibi Netanyahu. In the edited images, President Obama looked like Moe from the Three Stooges. It was hard not to laugh.

In fact, there was only one slide the president felt the need to change. A few weeks earlier, the White House had released a photograph of the president shooting clay pigeons at Camp David, and his critics accused him of doctoring the image. Obama, they insisted, was a gun grabber. His only reason for holding a firearm would be to melt it into a solar panel, or stuff the barrel with a gay pride flag. Their accusation was ludicrous, of course, utterly bananas. Naturally, it spread like wildfire on right-wing blogs.

Now, at Cody's suggestion, we were going to present an "undoctored" image to the world. In our picture of what had "really" happened, POTUS was still firing a gun. But in the background, we added a lightning storm, a monster truck, and a kitten the size of a black bear shooting lasers from its eyes. As we prepared to leave the Oval, POTUS held us back. He had an edit to request.

"Can we get a NASCAR in there?"

"We can do that," I said.

President Obama smiled contentedly. Then he perked up, struck by a sudden insight.

"Can Biden be driving the NASCAR?"

If only the president's enthusiasm was more infectious. In 2009, pitching jokes as an anonymous intern at a speechwriting firm, I had imagined the glories of responsibility. Now, four years later, I was learning the truth: responsibility kind of sucks. Every argument over a punch line chipped away at my psyche. Each time a member of the joke-writing diaspora schemed to get more of their own material into the script, it was my job to scheme back. Not a moment passed when it didn't feel as though everything was falling apart.

And yet in the world outside my head, everything was coming together. The slides looked good. The jokes were solid. Steven Spielberg, Tracy Morgan, and Barack Obama playing Daniel Day-Lewis playing Barack Obama had filmed their scenes. By Friday afternoon, just twenty-four hours before the dinner, I began to consider the possibility that everything might go as planned.

And then, sitting in my office, I got a phone call. It was Terry. He had a question.

"So I'm looking through these pictures of the president with the First Lady's bangs, and I'm just wondering, is the joke supposed to be that POTUS looks like Hitler?"

I immediately opened the slides in question The first photo was harmless. So was the second. Then I reached the third slide, the one with the president and the Israeli prime minister.

"Oh," I said.

"Yeah," he said.

It was shocking. The president didn't ordinarily look like Hitler in photographs. He certainly didn't look Hitler-y in person. But at that exact angle, with that specific haircut, there was no mistaking it. Even without the mustache, the resemblance was uncanny.

Just a month earlier, I might have tried to keep the slide in the script anyway. It was funny. Besides, would anyone think we were *trying* to make POTUS look like a Nazi? But I wasn't risking another international incident. I was done listening to Lips. Thanking Terry profusely, I hurried to save my speech.

On April 27, the day of the dinner, Cody was at a wedding. But Favs and Lovett were in town for the occasion, and they joined our day-of meeting instead. These final run-throughs were always casual. Instead of sitting at his desk or in his armchair, the president plopped down on a couch.

As usual, the meeting began with small talk. POTUS asked Favs about his new speechwriting business. He teased Lovett about his life in L.A. Meanwhile, I sat there stunned. I couldn't believe how casually my former colleagues charmed the president. What better proof that bringing me back to the White House was a mistake?

I was so busy wallowing that I barely heard POTUS ask a question.

"What happened to that picture of me and Bibi? I liked that one."

Favs jumped in. "We had to cut it."

"Well, why?"

Suddenly, the Oval Office fell completely silent. Plenty of people have compared the president to Hitler. But in all of American history, no one had ever compared the president to Hitler *to the president*. And none of us wanted to become the first.

It turns out that time slows down when you're trying not to insult the commander in chief. I remember considering, in surprising detail, just how doomed we were. Favs wasn't saying anything. Lovett wasn't saying anything. I wasn't saying anything. There was no way out.

There must be someone in this room who can tell the president the truth, I thought. But I couldn't begin to imagine who that might be. We needed someone bold. We needed someone daring.

We needed someone who didn't give a fuck.

In that moment, out of nowhere, I heard a voice. And it was Lips.

"I'm sorry, Mr. President," I heard myself say, "we just couldn't use that picture. You kind of look like Hitler in it."

The moment the words left my mouth, my out-of-body experience ended. What had I just done? All eyes were on POTUS. Nothing like this had ever happened before.

And then, President Obama began to laugh. Not his ordinary laugh, a self-aware one that was an act of judgment as much as reflex. This was an expression of something visceral inside him, a place beyond even his formidable self-control. He clasped his hands together. His feet kicked off the floor. He rocked back into the couch cushions. For just a fraction of a moment, I even think he forgot which person was the president. I had never seen him laugh so hard, and would never see him laugh so hard again.

Eventually the meeting returned to normal. Favs and Lovett resumed their confident banter, and I went back to sitting quietly on the couch. But I realized something. For the first time, I wasn't afraid.

The president finished his read-through not long after. We stood to leave, clutching our copies of the script. Before I could reach the door, however, POTUS looked right at me.

"Thanks, Litt," he said.

10

JUICE IN PURGATORY

A few hours later, I was backstage at the Correspondents' Dinner, washing my hands in the restroom reserved for POTUS. Suddenly, I heard the pounding of a fist. This was not the polite, inquisitive knock you associate with someone who hopes to use the facilities. It was the frantic, violent knock you associate with someone who hopes to flush cocaine.

I flung the door open to reveal an embarrassed-looking Secret Service agent. Thirty seconds earlier, he had grudgingly allowed me into the president's hold room. Now he regretted his mistake. For standing behind him, looking Zen-like by comparison, was President Obama.

I should have expected this. Who else would warrant such urgent knocking? But in my hurry, I had no time to think. All my brain could process was that I had opened a door and found an acquaintance on the other side.

"Oh, hi!" I said, as if Barack Obama were a second cousin

and not the most powerful person on earth. Fortunately, he seemed not to mind.

"Litt. We still funny?"

"Yeah, I think so."

Just a week earlier, I had worried my White House career was on the verge of an early end. Now, hustling out of the hold room, I could feel my head begin to swell. *I just said hello to the president, and Secret Service didn't shoot me. Maybe I'm becoming indispensable!*

If I ever needed to regain perspective, however—or be reminded how quickly things can change—all I had to do was see a movie. In 2013, several films were released in which the White House was blown to smithereens by paramilitaries, North Koreans, or people who simply disliked G.I. Joe.

I did not care for this trend. More than once I went to the movies to unwind after a stressful workday, only to see my office obliterated during the coming attractions. In one trailer, for *Olympus Has Fallen,* a plane bristling with machine guns strafed the State Dining Room. The rest of the audience barely noticed, but I clutched my popcorn and winced.

Oh, man, I thought. *Chase is totally fucked.*

It was discomforting to see my colleagues so casually exterminated. It was even more discomforting to realize my own death wasn't interesting enough to show on-screen. Cabinet secretaries were executed in dramatic fashion. Secret Service agents were launched skyward by explosions or cruelly double-crossed. But low-level speechwriters? We weren't even cannon fodder. Incinerated in a single CGI blast, my entire existence was mere kindling for the sweaty heroics of Gerard Butler or Jamie Foxx. During one trailer, as a grim-looking terrorist launched a rocket into the Grand Foyer, I found my rooting interest was with neither the president nor the hunky antihero destined to save him.

There's still hope! I thought. *Maybe the speechwriter's at the dentist.*

These movies, it must be said, were not entirely realistic. I never once worried that a shape-shifting supervillain would compromise the Oval Office before raining destruction upon earth from outer space. But the part about my life not being worth much? That was distressingly real. While junior-level staff rarely spoke about it, we knew the implicit bargain. We were low-profile people in one of the world's most high-profile targets. In a worst-case scenario, our final moments might be spent watching an escape pod rocket skyward and thinking, *Well, fair enough.* It didn't matter how many times I said hello to the president. I was completely disposable.

Knowing this, it was impossible not to envy certain colleagues. Terry, for example. He wrote national-security speeches, worked in the West Wing, and always wore a large silver key around his neck. Clearly he had access to a top secret bunker and would be one of the lucky few to repopulate the earth. If the earth ever needed repopulating, it could do worse than Terry. Still, it was hard not to feel left out.

Once you started looking, insecurity about security was everywhere. One morning a poster appeared outside Ike's:

ACTIVE SHOOTER PREPAREDNESS WORKSHOP

A few days later, I noticed another, smaller sign taped over the first: SECOND SESSION ADDED DUE TO OVERWHELMING INTEREST.

There was only one place where, in both the emotional and physical sense, I felt completely secure. The plane. From the day I arrived in 2011, I dreamed about tagging along on a POTUS trip. Now, as a presidential speechwriter, I was permitted to travel when remarks were delivered out of town.

As you've probably gathered by now, many aspects of working at the White House are not as cool as you would expect. Air Force One is *exactly* as cool as you would expect. I'll never forget my first drive to Andrews Air Force Base. Passing through the thick metal gates, our van pulled onto a tarmac the size of a golf course. We zipped by Cessnas, 757s, cargo carriers. Finally we stopped, just a dozen yards from the plane that dwarfed them all.

I had seen Air Force One before, of course. On TV. In movies. On the news. But to me, the presidential aircraft had always been like the surface of Jupiter or Art Garfunkel's living room. I knew such a place existed. I just never imagined setting foot there myself.

Now the plane's sky-blue belly was so close I could touch it. I didn't, of course. Instead, I walked carefully below a jet engine big enough to hold a trailer, and at the bottom of the rear staircase, I gave my name to a uniformed guard. He ushered me forward with a nod. Just a few steps later, I gasped. I was standing next to the exact same cargo hold where Harrison Ford killed his first terrorist in *Air Force One*.

Todd, a flight attendant, offered me a tour of the aircraft's five sections. Reporters sat at the very back of the plane. Next came the VIP guest cabin, which was followed by the staff cabin for people like me. After that came the conference room. This was where President Obama usually spent his time, either working or playing cards. Finally, at the nose of the aircraft, POTUS had a fully appointed office with a small private bedroom attached. Air Force One was (in the old, pre-Trump sense) a Russian nesting doll of access. You could always move backward from your assigned seat, but you needed permission to move up.

Not every part of the plane was fancy. With its leather recliners, wood trim, and wall-to-wall beige carpeting, the interior of the staff cabin resembled nothing so much as my grandparents' den. But what Air Force One lacked in luxury, it made up for in countless other ways. With airspace cleared ahead of us, we shaved hours off coast-to-coast trips. From the phone in my armrest I could dial anyone, anywhere. (President Obama spoke with foreign leaders. I called Jacqui to show off.) If I caught an ankle in the extendable footrest, or a forkful of crispy taco salad went down the wrong pipe, a pint-size emergency room stood waiting to receive me.

The plane's most extraordinary feature, however, was the image it projected. Nothing, not even the White House, so clearly symbolized the president's influence and reach. *We are America,* the aircraft seemed to say. *You know us. You envy us. You've seen us on TV. And the man up front can do anything.*

BUT COULD HE REALLY? IN A DIVIDED COUNTRY, WAS OBAMA'S PLACE in history still his to decide? To put it less poetically, could he make Washington work?

We certainly hoped so. We were sick of gridlock. Voters were, too. Yet where our economic plan was intensely detailed, our anti-gridlock plan was vague. Our theory—inasmuch as we had one—was that after a 2012 victory, obstruction would melt away.

"I believe that if we're successful in this election—when we're successful in this election—that the fever may break," POTUS declared.

Just a few weeks after Election Day, a tragedy and its aftermath seemed poised to prove him right. On December 14, at an elementary school in Newtown, Connecticut, a young man with an AR-15 rifle murdered twenty children and six adult staff. There had been mass shootings before, but this was different. No one, not even the president, could hear the news without feeling something rip apart inside. As POTUS addressed the nation from the briefing room, he paused, eyes watering, completely overcome.

The president's tears captured the country's attention, and rightly so. But no less important were his words at a prayer vigil a few days later, when he spoke about his own worries as a father.

"You realize," he said, "no matter how much you love these kids, you can't do it by yourself, that this job of keeping our children safe and teaching them well is something we can only do together."

Here at last, in the midst of unimaginable anguish, was Barack Obama's clearest argument for government's role. His was not a technocratic diagnosis, a blanket assumption that Washington knows best. Nor was it partisan: there was room for both Democrats and Republicans in the vision he described. Instead, his view of government was rooted in the responsibilities of family and the facts of modern life. In the twenty-first century, we cannot raise our kids alone.

For a moment, Washington seemed to agree. Senator Joe Manchin, a West Virginia Democrat, joined his Republican colleague Pat Toomey to try and strengthen firearms background checks. Their proposal wouldn't end gun violence completely. Still, it was an obvious start. And it was popular; even a majority of Republicans supported the idea. If I had learned anything from watching *Schoolhouse Rock!*, this bill was destined to pass.

But the National Rifle Association understood something I didn't. *Schoolhouse Rock!* was a lie. In Mitch McConnell's Congress, even the most appealing bills were at the mercy of nearly insurmountable math. For starters, advancing any piece of legislation took sixty votes in the Senate—a three-fifths majority. If that doesn't sound daunting, consider this: no presidential candidate has won a three-fifths majority of the vote since Richard Nixon in 1972.

It was easy, in other words, to make nothing happen. And

on the background-check bill, nothing did. The NRA held its ground. Its allies in Congress quietly ignored popular opinion. The proposal died with a whimper instead of a bang.

Not that passing the Senate would have done the bill much good. After the 2010 landslide, Republicans redrew House districts to protect incumbents from pesky annoyances like the voters' mood. Such rampant gerrymandering was a disgrace. It was also a success. In 2012, Democrats won a slim majority of votes for Congress. But Republicans won a large majority of the seats. An election that was supposed to teach lawmakers to respect the will of the people had instead taught them the opposite. Our representatives had nothing to fear.

No wonder, then, that the fever didn't break. If anything, it got worse. And just as McConnell predicted, people blamed Obama for the dysfunction others caused. The Tuesday after the Correspondents' Dinner, POTUS held a press conference to mark the first hundred days of his second term. It was supposed to be a victory lap. Instead, ABC News correspondent Jon Karl ticked off a list of laws Congress had failed to pass. Then he delivered his challenge.

"My question to you is: Do you still have the juice?"

Other presidents might have relished this moment. Bill Clinton, for example, loved the drama of the nation's highest office. Nothing made him happier than playing himself in the movie version of his life.

But POTUS wasn't like that. Where Clinton's ego expressed itself in an insatiable craving for power, Obama's expressed itself in the absolute conviction he deserved it. This faith in his own virtue is what made him authentic. He handwrote thoughtful responses to letters from conservatives. He met with kids from Make-A-Wish without mentioning it to the press. But the president's high-mindedness also made him impatient when he felt other, less-worthy politicians were being let off the hook.

"Jonathan"—it was as if the reporter were four years old, and the president was explaining, for the thousandth time, why they couldn't have ice cream for dinner. "Jonathan, you suggest that somehow these folks over there have no responsibilities, and that my job is to somehow get them to behave. That's *their* job. Members of Congress are elected in order to do what's right."

POTUS wasn't incorrect. McConnell's behavior *was* reprehensible. The do-nothing Congress *was* holding America back. But just because Washington is contemptible doesn't mean a president benefits from showing contempt. We heard the same thing in focus group after focus group. Americans empathized with their commander in chief. They knew he had an impossible job. But they did not, under any circumstances, want to see him complain.

I, however, am not the president, and will take this opportunity to complain bitterly. Jon Karl's question was ridiculous. Not just useless, but harmful. The presidency is not an orange. You could have put Mitch McConnell on a fast, a diet, a weeklong cleanse: no amount of juice was going to change his mind. He believed he could elect more Republicans by constantly opposing President Obama than by occasionally joining him. He believed this, in part, because reporters who could help Americans understand the true cause of gridlock would rather ask POTUS about juice. Thus, obstruction continued. Nothing got done.

Actually, that last part is not quite fair. While no new laws got passed, outside Washington, real change was taking root. The economy, on the edge of collapse four years earlier, was steadily creating jobs. Clean energy production rose. Foreclosure rates fell. So did deficits, in part because George W. Bush's tax cuts for the superrich expired at the beginning of the year.

During my first few months back at the White House, these signs of progress were at the heart of our communications strategy. If Washington represented America at its worst, we would highlight America at its best.

Which is why, for Memorial Day 2013, POTUS scheduled a visit to the Jersey Shore. Less than a year earlier, Hurricane Sandy had devastated the area. Now, the tight-knit community was bouncing back. For America, the shore was a model of resilience, a shining example of our country's determination and grit. For me, however, the shore was important for a different reason: Jacqui's parents lived there. In fact, their home was just a few minutes from the Asbury Park Convention Hall, where the speech would take place. I could live a thousand years and never see a better point-scoring opportunity.

Naturally, I volunteered to write the remarks.

May 27, the morning of the speech, brought the exact opposite of beach weather. It was cold and rainy, with suffocating fog. But I wasn't worried about sunshine. I was worried about upgrading Jacqui's mom and dad to VIP seats. As I watched Chris Christie win a stuffed bear for POTUS on the boardwalk, I mentioned my predicament to Bobby, one of the president's aides.

"Why don't we just put them in the photo line?" he asked.

I hadn't even thought to request this. Unlike tickets to a speech, which could be scattered to staff like confetti, a picture with the president seemed impossible to obtain. But on the road, the tear-inducing onion of White House bureaucracy shed its layers. There was no obscure form requiring five signatures, no office that could grant my request but was only reachable via fax. Bobby gave me a name. I sent an e-mail. An hour later I was standing backstage at the convention hall, happy to be out of the rain and waiting for POTUS to begin his remarks.

"Your girlfriend's here?" asked the prompter operator.

"Yeah," I said. "I got her and her parents in the photo line."

For a moment, he didn't say anything. Then he smiled, caught somewhere between jealous and impressed.

"You're gonna have to work real hard to screw that one up."

As if the day wasn't perfect enough already, we got to take choppers back to the National Guard base where Air Force One was parked. For these short hops, the president rode in Marine One, the champagne of helicopters. Junior staff, however, were herded onto Chinooks, twin-rotored military transports that look like flying RVs. Windows were open to the elements. Two long benches took the place of seats. The interior smelled like a blend of motor oil and locker room. It was paradise. More than walking through security checkpoints, or even flying on the presidential plane, the helos made me feel badass.

In fairness, not everyone on these flights needed help with their badassery. While I admired the view, a half dozen of my seatmates adjusted the helmets covering their crew cuts. As I extended my phone for a helicopter selfie, they cradled assault rifles in soft cases on their laps. The pockets of their bulletproof vests bulged with deadly knickknacks. The glassy look in their eyes suggested a decidedly informal approach to human life.

They were members of the Counter Assault Team, or "CAT team" as it was more commonly known. I once asked a more experienced speechwriter what separated these people from the rest of Secret Service. Here was his reply:

"If something bad happens, Secret Service gets the president out of trouble. The CAT team finds the trouble and kills it."

That afternoon on the shore, as we boarded the helicopters, the CAT team members assumed their typical posture. Backs straight. Chins up. Mouths flattened into an expression best described as "last thing you see before you die." I, however, was less stoic. The rain was still falling heavily. The fog had only gotten thicker. I was surprised we were cleared to fly. As we lurched upward, I remembered someone telling me that, rather than using radar, our pilots navigated by sight.

It's okay, I told myself, *there's a grown-up in charge of the weather call. They won't let us do anything unsafe.*

Twenty minutes into our ten-minute flight, I reconsidered.

For those who have never been suspended midair inside a thundering Chinook surrounded by smothering fog, let me try to explain. Imagine a kind of military-grade sensory deprivation chamber. The rotors are the world's loudest white noise machine. Out the window lies ceaseless, unchanging gray. As cold air floods the cabin, numbness sets in.

If you're like me, this experience will lead you to cycle through the following questions: Are we flying in the right direction? Are we flying in any direction? Are we dying? Are we already dead? Is this limbo? Is this purgatory? Trust me: there are far more pleasant ways to spend twenty minutes on a Monday afternoon.

Or thirty minutes. Or forty.

We didn't crash, of course, or you wouldn't be reading this. But by the time our helicopter landed, we had spent nearly an hour in the air. And it turned out we really were in danger. Once the choppers were airborne, the pilots realized visibility was practically zero. There was a genuine possibility that two helicopters would blindly bump each other, and that one would contain the president of the United States. That's why our flight took so long. Rather than assemble in tight formation and risk a crash, the helos had lined up single file. When it felt like we weren't going anywhere, we *weren't* going anywhere. Our chopper was waiting its turn to land, without instruments, in pea-soup fog.

As you can imagine, there was plenty of nervous chatter on the flight home. But for the most part, I didn't join in. Sitting silently in my staff-cabin recliner, I felt less frightened than foolish. How many times would I need to relearn this lesson? No one is all knowing. No one is infallible. Not the speechwriters, not the people making the weather call, not the president of the United States.

Even the trained killers of the CAT team were only human. Somewhere above New Jersey, deep in foggy purgatory, our helo had hit an unexpected patch of turbulence. I distinctly saw two of them flinch.

RETURNING TO YOUR COLLEGE CAMPUS AND BRAGGING ABOUT YOUR recent Air Force One flight is not the only way to feel good about yourself. That said, it works.

I learned this shortly after my near-death helicopter experience, when I attended my five-year reunion. A half decade out of school, my friends were asking the questions confronted by all twenty-somethings. What fulfills me? Are my goals worthwhile? How do I define success? Working at the White House meant I could skip these questions entirely. I once compared the president to Hitler during a meeting. How much more fulfilling can life get!

"You're living the dream," I was told, at least a dozen times.

What I didn't tell my classmates was that, while my higher purpose seemed certain, my moment-to-moment well-being was in constant flux. I was living not one dream, but two. In the first, I flew around on Barack Obama's private jet while he helped me score brownie points with my girlfriend's parents. It was awesome. But the second dream was more fevered, more troubling. I was hovering midair. Men with guns surrounded me. I didn't know where we were going. The pilots were flying blind.

Was I in paradise, or in limbo? In a state of profound gratitude, or persistent unease? At the White House, it was often hard to say.

At least I was no longer getting blown up quite so regularly. Still, I continued to open Cody's edits as though their contents might explode. By now, I didn't even have to check tracked changes to know what would befall my ego. Like most EEOB employees, I studied the West Wing with anthropological intensity, and had learned to translate my boss's unique dialect of one-line e-mails.

My edits. *Unmitigated disaster. Pure garbage. Rewrite.*
Here are my edits. *I disliked this, but I didn't completely hate it.*
Some edits. *This was acceptable, but only by the smallest possible margin.*
Good job. *Good job.*

The final category of e-mail, and by far the most precious, was any message containing the words *boom!* or *bro*. These were special. They meant you were totally killing it and had established yourself as a valued member of the team.

Four months after my return to the White House, however, *booms* and *bros* remained elusive. While the Correspondents' Dinner had been a hit, more serious speeches continued to frustrate me. I wasn't killing it at all.

Frankly, neither was POTUS. The reelect was a hearty "good job" from voters, but by the summer of 2013, he was firmly in here-are-my-edits territory. Nor was there any reason to think his approvals would rise. With the fever raging, and our legislative agenda stalled, Americans were losing faith. Fog was enveloping the second term. We needed something to break through.

That something was a speech. The decision was made to return to Knox College, the site of POTUS's place-in-history commencement. For a week, Cody barricaded himself in his office. When he emerged, it was with a set of remarks entitled "A Better Bargain for the Middle Class."

No one thought the better-bargain address would boost our approvals. Thanks to a splintered bully pulpit, even presidents can no longer tell America what to think. They can, however, tell America what to think *about*. That's what Cody's speech was designed to do. Sometime in the 1970s, the remarks reminded us, higher productivity stopped generating higher wages. The cause-and-effect relationship between work and reward had disappeared. Restoring this link would be the great project of President Obama's second term. Nor did POTUS stop there. In the rest of his address, he described what he called "cornerstones" of middle-class life. Good jobs. Affordable health care. Economic opportunity. A chance to buy a home. A secure retirement.

This was riskier than it sounds. There's a widespread belief that most politicians don't try to keep their promises. That's false. They do. But it's for precisely this reason that politicians generally avoid promising at all. Now, POTUS had laid out five separate standards by which to judge his performance. It wasn't hard to see the subtext. *If I can deliver on these priorities, I will earn my place in history. If not, I won't.*

AS EXPECTED, PRESIDENT OBAMA'S GUTSY NEW TONE DIDN'T reverse the effects of gridlock. But it did lift morale. For the first time in months, there was hope that the fever could be sidestepped, that our success or failure would not be left to Congress to decide. Our supporters took note.

"He was good!" friends and family told me, assuming, as they often did, that I had written the remarks.

"You should have seen the first draft," I said, trying to sound knowledgeable without actually claiming credit. "It was a real Christmas tree."

They had no idea what I was talking about. That, of course, was the point. Like every workplace, Obamaworld was home to a members-only vocabulary. The more disposable we felt, the more comforting it became to rely on phrases outsiders didn't understand.

Christmas Tree. *Noun.* A speech hopelessly weighed down with personal agendas and irrelevant policy details.

Due-Outs. *Plural noun.* Tasks assigned at the end of a meeting.

Bigfoot. *Verb.* To pull rank on. ("Sorry I had to bigfoot you, but that conference call was for senior staff.")

Click. *Noun.* A picture at a photo line. ("Shouldn't take long. Only twenty clicks.")

Real Person. *Noun.* An American living outside Washington who is not famous and does not work in government. Often shortened to "RP."

Equities. *Plural noun.* Interests.

Stakeholders. *Plural noun.* People with equities.

Socialize. *Verb.* To circulate a policy informally. ("Let's socialize this with stakeholders. We don't want to ignore any equities here.")

Along with the jargon came shorthand. Acknowledgments in speeches were *acks.* The apparatus POTUS read from was *the prompter.* His big annual address to Congress was *the SOTU.* Perhaps most important, the metal flying thing that carried the president around was always *the plane.* Nothing exposed you as a White House newbie faster than saying "Air Force One."

Then there was the shorthand for Obama himself. In 2011, *POTUS* was still a cool-kids-only word. Then everyone started using *POTUS,* and the cool kids switched to just *P.* The initial proved even better than the acronym; on a BlackBerry, it required only one thumb to type. It also implied you were far too busy to waste time on four extra letters. Every so often, someone took the opposite approach, referring to the president as "the president." But this was done rarely, and only as a power move.

"POTUS sent edits."

"P is running late."

"The president personally approved this. Do you still want to change it?"

At times, White House etiquette was no less complicated than vocabulary. If someone forwarded an e-mail without additional text in the body, for example, it was always possible they were just passing on information. More likely, however, they were sending a message in code:

"I don't want an archived record of me saying so, but this is the stupidest fucking thing I have ever seen."

There was one final appendix to the Obama glossary: terms that signaled the absence, rather than the presence, of the thing they described. Chief among these was *the process.* I'm not sure if they used the term in the West Wing. In the EEOB, however, *the process* referred to the mysterious black hole all ideas entered and few escaped. POTUS almost shot a video for Comic-Con, but the process moved slowly and it got pulled down. When I suggested a *Hamilton* parody video featuring the president, I doubted it would survive the process, and it didn't.

It was the process that determined which speeches opened with an engaging anecdote and which began with a sleep-inducing parade of acks. The process even dictated where and when POTUS would speak. "How does he decide what to talk about?" interns would ask. I would dutifully pretend to know. But really, their guess was as good as mine. Maybe topics were decided upon in a scheduling meeting. Maybe they were delivered in bundles by storks.

All I knew for certain was that, just as a surplus of stakeholders led to a Christmas tree, a big speech like the better-bargain address led to a series of message events. These were the bread and butter of POTUS speechwriting, the rhetorical equivalent of routine maintenance on your car. The goal was not to land in the history books. It was to focus America's headline writers on a specific place and issue.

OBAMA VISITS MEMPHIS, LAYS OUT PLAN FOR COMMUNITY COLLEGE

IN KANSAS CITY, PRESIDENT PRAISES COMEBACK IN MANUFACTURING

In late July, it was decided that POTUS would do one message event for each of his middle-class cornerstones, and that I would take housing policy. This was hardly a vote of confidence. Health care was timely. Education was inspiring. Housing was a sedative. Handing me remarks about mortgage rates and home equity lines of credit was like sending the chubby kid to right field.

Like any good right fielder, however, I pretended not to notice. Just as I had done before all my speeches, I leaned back in my ergonomic chair, my dress shoes adding to the collage of scuff marks on my wall. I relished this moment. With nothing but a blank screen and a blinking cursor, I was Beethoven sitting down at his piano. I was Picasso picking up a brush.

Then, suddenly, an epiphany: I wasn't doing my job at all. My work wasn't art. It was craft. For all but the tiniest handful, government service is about competence, not genius; precision, not brilliance. It was time to stop pretending the process didn't apply to me.

So that's what I did. Instead of reinventing the wheel, I harvested bits and pieces of old language, gluing them together with lines from Cody's Knox College remarks. I punched up old anecdotes instead of finding completely new ones. True, I added a few new flourishes: a joke here, an RP story there. But there was no getting around it. It was the least original work I had ever produced.

As if that wasn't process-y enough, I had to run it by the fact-checkers. The idea of a White House that checks facts—the idea of a White House that *believes* in facts—already seems like a relic of another time. Yet less than six months ago as I write this, there was an entire research office responsible for making sure that the president's statements were true. Our researchers saved me from countless embarrassments. They were invaluable, not just to the president but to democracy itself. Hardworking and unfailingly humble, they were model public servants. Also, they bothered the shit out of me.

It wasn't personal. The research office was just a few steps away from the speechwriting office. We were on the same team and friendly outside work. Inside the building, however, we feuded as bitterly as the Sharks and Jets.

The problem, as far as researchers saw it, was that any speechwriter left unsupervised would begin writing fiction. The problem, as far as speechwriters saw it, was that researchers were almost comically risk averse. Terrified of even the slightest error, they highlighted line after line in yellow, with little notes underneath. These comments could make me bang my head against a wall.

America is the greatest country on earth.

Actually, the Nordic countries surpass us on several key measures.

Our economic policies are working.

Just flagging that Republicans might not agree.

I didn't blame the fact-checkers for going a little overboard. In their world, a surplus of caution was a surefire way to get ahead. But the result was an eight-year-long, cover-your-ass arms race. Research began adding *nearly* or *almost* to every assertion. POTUS began skipping over the words *nearly* and *almost* every time he spoke. Research began deferring to policy on even the most obvious questions. Speechwriting began suggesting that if every question was so important, perhaps researchers should, well, research a few answers themselves.

I didn't often collaborate with my fellow speechwriters on lines or paragraphs. But in the EEOB, we swapped anti-research tactics the way allies shared intelligence during the Cold War. We had no choice. If research had gone over the paragraphs you've just finished reading, here are just a few of the notes I would have received.

Just a few steps away from

Flagging that stride length varies by person, sometimes by large amounts.

We were on the same team.

Defer to the Office of Presidential Personnel.

. . . as bitterly as the Sharks and Jets.

At the end of West Side Story, someone dies. Can we say "almost as bitterly"?

POTUS began skipping over the words *nearly* and *almost* every time he spoke.

Should change to "nearly every time he spoke."

We had no choice.

Do you have backup proving this? If not, "We had limited choices" would be safer.

For all its truth-preserving merits, research was the process personified. No matter how excited I was at the outset, it left me completely, totally drained. My housing speech was scheduled for Tuesday, August 6. By the time I sent a draft to Cody, that

Sunday afternoon, I never wanted to see it again. To be honest, I never wanted to write another speech again. Every ounce of emotional energy had evaporated. Every drop of joy was gone.

Exhausted and famished, I went to lunch with Jacqui at a Thai restaurant near my apartment. Before we could even order, my BlackBerry vibrated. It was Cody. Too tired to care anymore, I opened the e-mail from my boss.

Great job, bro!

Just like that, the exhaustion vanished. I knew my next set of remarks would be another roller coaster. So would the set after that, and the set after that. It didn't matter. I couldn't wait to dive back in.

This, I was realizing, is what it really means to work at the White House. Ping-ponging between emotional extremes, I had finally arrived at my inner common ground. I was in paradise *and* limbo. Indispensable *and* disposable. Defined by process *and* purpose. Washington was in the grip of unbreakable fever, yet there was nowhere I'd rather be.

Was my job as wonderful as I'd imagined when I'd first walked through the gates? Of course not. But it was also more than enough. On the day of the housing speech, as I sauntered into the staff cabin and snatched a paper card off my seat, I might even have told you I was living the dream.

Mr. LITT, the card said. *Welcome aboard Air Force One.*

11

THE HOLY WAR

"We already have a Christmas," I told Jacqui. "It's called Yom Kippur."

This was during our first July together, when she was giddy about being halfway to December and I was not. To be clear, I didn't *hate* Christmas. I had no desire to be a Grinch. I simply placed Christmas in a category with *The Big Bang Theory,* the Dave Matthews Band, and peyote. It wasn't my thing.

For Jacqui, this was unacceptable. Presbyterian on her father's side, Catholic on her mother's, she was not especially fervent about most aspects of her faith. But when it came to Santa Claus, indoor trees, and *It's a Wonderful Life,* her evangelism easily surpassed Saint John's. That I did not accept Christ as my lord and savior left her unbothered. That I did not accept Christmas as the most wonderful time of the year made her fear for my very soul.

Hence the inquisition that first July. "What do you do on *your* Christmas?" she demanded.

"Well, for starters, we don't eat anything." I hoped to sound high-minded, even spiritual. But Jacqui remained unconvinced.

"If you don't eat all day, then what do you do?"

"Mostly we pray," I said.

"For what?" she said. "For food?"

IF MY GREAT-GREAT-GRANDFATHER WERE BROUGHT BACK TO LIFE, imported from eastern Europe, and told this story, he would surely be furious to hear I was dating outside my religion. But his disapproval wouldn't last long. One look at two dudes kissing in the gayborhood near my apartment and he would drop dead of a heart attack on the spot.

In this way, Jacqui and I were a product of relaxing attitudes all over. Interracial, same-sex, interfaith. Second marriages, co-habitation, no-fault divorce. The overall trend, decades in the making, was neatly summarized by a leading gay-marriage slogan.

"Love is love."

There was just one big exception. Politics. In 1960, a mere 5 percent of Americans said they would be upset if their child married someone from the opposite party. A half century later, that number had exploded eightfold, to 40 percent. Partisanship, in other words, was playing a dividing-line role once reserved for religion and race.

As a loyal Democrat, I felt firsthand the effects of political tribalism. In the mid-2000s, for example, I knew President Bush spent a scandalous amount of time on vacation. This was not a conclusion I reached via analysis. It was dogma. Then I got to the White House and learned the truth: presidents are never really on vacation. No matter how many days he spent clearing brush at his Texas ranch, George W. Bush was permanently at work. This was a mild awakening, less "God is dead" than "God has a peanut allergy." Still, it came as a shock.

Yet that is nothing compared to what happened across the aisle. Yes, it's true that both Democrats and Republicans became polarized in the early twenty-first century. But it's also true that both *Porky's* and *Psycho* depict bad manners in the shower. Degree counts.

After all, when Obama took office, postpartisanship was kind of his thing. On big issues—education, climate change, health care—he borrowed ideas from Republicans. Rather than starting from one extreme and negotiating toward the center, his early proposals often arrived with compromise baked in. A few decades earlier, these gestures might have been reciprocated. But this was the age of the Tea Party. Each time Obama entered new common ground, a kind of white flight occurred.

By 2013, thanks to backlash against a Spock-like president, even logic itself had become partisan. More and more, the Republican Party was defined not by arguments but by articles of faith: Climate change wasn't real. Voter fraud was rampant. Deficits were rising instead of falling. More guns meant less gun violence. President Obama's economic recovery plan had yet to create a single job.

That these ideas were demonstrably untrue was not, in and of itself, a problem. My own religion forbids consuming milk with meat. As anyone who has ever eaten a cheeseburger can tell you, this makes no sense. And that's fine! If every tenet was rational, we wouldn't call it a faith. But journalists who continued to treat the GOP as a traditional mesh of interest groups invariably tied themselves in knots. The conservative movement had undergone a transformation. The Republican Party had become a kind of church.

This did not mean its members were in complete agreement. As with any self-respecting religious institution, a million sects and subgroups vied for control. It's impossible to classify every denomination. To understand President Obama's second term, however, all you need to know are the following three:

First, the Country Clubbers. Guardians of the GOP's upper-crust traditions, they believed in lower taxes, less regulation, and being polite. They were led in Congress by Speaker John Boehner. They held out hope for the resurrection of Mitt Romney. Their fortunes were not on the rise.

Second, the Flat Earth Society, with Sarah Palin as its patron saint. These were the hard-core conspiracy theorists. They insisted that President Obama had faked his long-form birth certificate. They were certain that bike-share programs were a world-domination plot fostered by the UN.

Finally, the Holy Warriors. Some of these crusaders were, in fact, religious. Others were more likely to quote *The Lord of the Rings* than Matthew or Luke. But regardless of where they spent their Sundays, what they shared was a worldview. Where traditional Republicans saw a debate between liberal and conservative, Holy Warriors saw an existential battle between good and evil. They warned endlessly of appeasement. They spoke of "defeating the Left" as though Satan's minions were amassed along the Pacific coast.

The Holy Warriors pursued Romneyite goals with Palinite fervor. For this reason, they were ascendant in 2013.

If we're being honest, however, what really put butts in pews was Obama. Nothing united Republicans more successfully than dislike of the commander in chief. And nothing provoked them more than his greatest legislative achievement. To Country Clubbers, the Affordable Care Act was a wealth transfer from rich to poor. To Flat Earthers, Obamacare meant death panels and government takeovers. To Holy Warriors it was the ultimate triumph of leftism—the final step on the road to Mordor.

As you might imagine, the Obama White House disputed

its signature law's evilness. When it came to the law's importance, however, we couldn't have agreed more. We were excited by all that Obamacare could accomplish: insuring millions, slowing the rise in health care costs, ending the ban on preexisting conditions. But our attachment went beyond cost-benefit analysis. The most sacred piece of Democratic orthodoxy was that government could improve people's lives. The most sacred piece of Republican orthodoxy was that it could not. If Obamacare worked the way it was supposed to, the debate would be over. There would be no doubt which was the one true church.

For this reason, both sides looked toward October 1, 2013, as a kind of Judgment Day. For at that fateful moment, Americans would be able to use something called Healthcare.gov to shop for insurance online.

On the surface, this didn't sound very exciting. A store! On the Internet! Designed by Uncle Sam! But the moment the marketplace launched, millions of people would be newly able to buy insurance. Once that happened, repealing Obamacare would mean robbing voters in every state and district of health care they could finally afford. For the law's opponents, Healthcare.gov was a doomsday device. It had to be stopped.

Enter Ted Cruz. On paper, the first-term Republican senator seemed an unlikely high priest for the Holy Warriors. According to his freshman-year college roommate, young Ted was so reviled he was said to leave a layer of "cruhz" (rhymes with *scuzz)* on every surface he touched. But in a party that saw no difference between compromise and betrayal, there were benefits to being bad at making friends. Before long, the jowly Texan had placed himself at the helm of a great crusade. The strategy was simple. If President Obama refused to defund his own health care law before October 1, Republicans would shut the entire government down.

The Country Clubbers whined and tut-tutted, but there was nothing they could do. To protect themselves from Tea Party challengers in their primaries, they had promised drastic action. Now the bill was due. Besides, the 2011 debt ceiling crisis left Republicans convinced that hostage-taking worked. Threaten to harm America badly enough, and President Obama would cave. The idea of a shutdown started out as crazy talk. By September, it was a near certainty.

What followed in the White House was a kind of white-collar doomsday prep. First, employees were divided into two groups. If your job was crucial to national security, or you were senior staff, you belonged to "essential personnel." I was part of the second category, and if I had been in charge of naming it, I would have been more polite. "Valued personnel." "Still-special personnel." Even just "Group Two." But the federal bureaucracy did not care about my feelings. By decree of the United States Government, I was officially nonessential.

After the culling came the stern legal warnings. During the shutdown, if I sent even a single e-mail from my BlackBerry, I could face a five-thousand-dollar fine. Unsure when I'd be allowed back in my office, I riffled through my desk drawers, disposing of perishable snacks. Last-minute errands in the West Wing were more serious. Staff assistants scrambled to teach their bosses to transfer a call or to choose a printer from the drop-down menu.

Finally, on September 30, President Obama appeared on television like a principal before a nationwide snow day, listing buildings and services that would close. A few hours after his announcement, I left work. I hung my badge deep in the back of my closet. Then, just to avoid temptation, I popped open my BlackBerry, yanked out the battery, and shoved both objects deep into a drawer.

For the next few days, I felt like a third grader whose school has burned down. On an abstract level, I knew the shutdown was hurting people. But on a personal level: vacation!

I wasn't alone. Washington's bars and restaurants have always catered to functional alcoholics coping with stressful jobs. With the jobs suspended, we were free to practice our functional alcoholism full time. At a neighborhood bar called the Brixton, four dollars would buy me a cup of something called Furlough Punch. The next morning, I could nurse my hangover at the Daily Dish, where federal employees got free coffee and congressmen paid double. After yet another all-day happy hour at Lou's City Bar, I could tuck into the "Boneless Chicken, Spineless Congress" special at Nando's or free nachos at Mango Mike's.

The fun lasted a week. After that, I became overwhelmingly bored. And worried. The shutdown was the biggest political fight in generations. It wasn't clear who would win.

Republicans tried everything to sway public opinion. Borrowing a page from the Country Clubbers' playbook, Fox News rebranded the event a "government slimdown," as if Uncle Sam were cutting carbs. When no one bought that, Palinites began buzzing about a new grassroots movement, "Truckers Ride for the Constitution." Organizers painted an epic picture: ten thousand tractor trailers, each driven by a pissed-off patriot, circling the beltway until Obama resigned in disgrace.

"Truckers will lead the path to saving our country if every American rides with them!" their Facebook page proclaimed.

I disagreed with their politics, but what really bugged me was their diction. Paths are fixed objects. You can't *lead* them anywhere. Not that it mattered. In the end, only a few dozen truckers showed up. They honked around a bit, inconvenienced a few commuters, and then led the path back home.

More than any presidential speech, the shutdown forced voters to notice which party was behaving like a bunch of children and which was not. Republicans' approval ratings began to plummet. With increasing desperation, they turned to the Holy Warriors. Surely Ted Cruz could save the day!

But this was where politics-as-religion fell short. Blinded by devotion, the faithful hadn't noticed that their prophet lacked a plan. The crusaders' strategy was best summarized by one of Cruz's close allies, an Indiana congressman named Marlin Stutzman.

"We're not going to be disrespected," Stutzman told the *Washington Examiner*. "We have to get something out of this. And I don't know what that even is."

As the shutdown dragged on, there was one final hope for the Republican Party. Surely Obama would cave. But not this time. In 2011, POTUS had paid dearly for negotiating with hostage takers. Two years later, he refused to repeat his mistake. His message was as clear as his opponents' was muddled. Stop hurting the country. Release the hostages. Or face the voters' wrath. On October 16, 2013, the shutdown battle ended in a rout. Republicans agreed to reopen the government. In exchange, Democrats agreed to let them.

The next morning I returned to the White House. Passing through the Secret Service checkpoint, I saw Denis McDonough, our new White House chief of staff, standing on West Executive Avenue. He was beaming as he shook everyone's hand.

"Welcome back! Good to see you. Good to see you. Welcome back!"

And why not celebrate? For five years, we had fought an endless string of uphill battles. Now, at last we were victorious. The promised land was finally in sight.

THE FOLLOWING AFTERNOON, I ALMOST SET BARACK OBAMA'S HAIR on fire.

In fairness, it was technically a team effort, one that began when Hope Hall decided to film the weekly address outdoors. By positioning the camera on the first-floor balcony of the residence, she could capture the immaculate green sweep of the

South Lawn. In the distance, across Constitution Avenue, the west face of the Washington Monument would glisten in the afternoon sun. What better way to show America that the government had reopened?

Once Hope laid out her vision, our A/V team began bringing it to life. In the White House, while civilians do the filming and writing, lights and sound equipment are handled by enlisted military personnel. I enjoyed spending time with the A/V guys, not least because our paths to 1600 Pennsylvania were often quite different. At age nineteen, for example, I was disciplined for falling short of a science requirement. Jared was disciplined for machine-gunning a dolphin during an offshore training exercise. Still, less than a decade later, his professionalism was a tribute to the reforming power of military service. Like all members of his team, Jared worked with an attention to detail I never could have matched.

For the weekly address the day after the shutdown, the person in charge of equipment was a skinny, muscly marine named Joe. Joe was a firm believer in contingency plans. Arriving at the taping, I found extra-long extension cords snaking into the residence. The spare camera battery was charging on its stand. A boom mic warded off ambient noise.

Joe even remembered to bring "diva lights," the suitcase-size LED arrays you sometimes see in behind-the-scenes pictures of models on a shoot. One high-wattage panel was positioned on either side of the president's chair. Then, to prevent any glare, Joe shielded each light with something called a diffusion cover, essentially a large Tupperware lid.

The taping was scheduled for early afternoon, so in theory the extra lighting was unnecessary. But we were lucky Joe played it safe. It soon became clear that even by White House standards POTUS was running way behind. I puttered around the residence for hours, killing time staring at portraits of Ben Franklin (pensive), and of Martin Van Buren's daughter-in-law (unexpectedly hot).

Finally, as the sun was beginning to set, we got word: P was moving. In anticipation, the diva lights burst on. But instead of heading to the balcony, the president and Denis, the chief of staff, began pacing the perimeter of the South Lawn. One lap. Two laps. Then three. It was nearly dark by the time they finished. Stepping out onto the balcony, President Obama looked tired and perturbed.

Then he sat down, and his attitude immediately shifted. I was always impressed by how rapidly POTUS cycled between moods at these tapings. He could arrive from a deadly serious meeting on Afghanistan, perk up to record a Michael Jordan birthday greeting, and then let his face fall for a solemn tribute to our troops. For weekly addresses, POTUS adopted a tone that was both formal and slightly severe, as if he were narrating a video about the dangers of backyard trampolines.

"This week, because Democrats and responsible Republicans came together, the government was reopened."

All of a sudden, I smelled something. Bug spray? Sunscreen? I glanced at Hope and Joe to see if they had noticed it, too. They hadn't. Probably a false alarm.

"Specifically, there are three places where I believe that Democrats and Republicans can work together right away," POTUS continued. He dropped both hands in an emphatic gesture, underscoring his point.

That's when I saw it.

The plastic lid on one of the diva lights had begun to smolder. A tiny, molten hole was releasing a curl of toxic smoke just a few inches from the president's left ear. I turned my head toward Joe. Just a few seconds earlier, his face had been reassuringly calm. Now, he looked constipated with fright.

But POTUS was on a roll, and he didn't catch Joe's expression. Nor, thanks to the breeze that evening, did he notice the sheet of burning poison by his head. "First, we should sit down and pursue a balanced approach to a responsible budget," the president said. His demeanor was cool and collected, a sharp contrast to the red-hot, glowing chaos nearby. Fumes began to build. The plastic took on the urgent, quivering quality of newspaper in a fireplace moments after a match is lit.

So naturally, without hesitation, I jumped in and saved the day. I extinguished the flame, rescued the president, and earned myself the gratitude of Barack Obama, the nation, and the world.

At least, that's what happened in my robust fantasy life. Here's what I did in my real life. Nothing. I remained totally frozen. Completely silent. And since then, I've asked myself the same question at least a thousand times.

Why? Why didn't I speak up? For that matter, why didn't Hope? Why didn't Joe?

The more I've thought about it, the more certain I am that our tongues weren't tied by shock or fear. What held us back was the same bureaucratic angel that saved me at the Correspondents' Dinner two years before. *Staaaaay in your laaaaaane.* I was in charge of writing the script. Joe was in charge of preparing the set. Hope was in charge of operating the camera. What we needed was someone in charge of making sure the president's head was not incinerated. But no such person existed. No one had been assigned that due-out. And in Obamaworld, straying outside your lane was a mortal sin.

Not a moment too soon, the president noticed Joe and me share a look of abject horror. Instantly, his attitude changed.

"Guys, what's going on?"

Silence.

"What?" he snapped.

Remarkably, we still said nothing. But while I remained paralyzed, POTUS's question snapped Joe to life. Grabbing a towel, he sprinted past me and smothered the diva light.

Once President Obama realized what had happened, he fixed us with a withering look. I prepared myself for a completely justified tirade. But to my surprise, his expression grew weary instead of angry. With a new plastic cover in place, he raced through the taping and into his house.

"Have a great weekend, everyone."

"Um, you, too, sir."

I am not saying Barack Obama was a saint, capable of forgiving any trespass. Here's what I'm suggesting instead: his responsibilities were so great, and he took them so seriously, that his employees' almost burning his face off barely registered on his list of concerns. There was too much else to worry about. The shutdown had wreaked havoc on our economy. The Holy Warriors were vowing to continue their crusade.

And finally, there was the issue that threatened Obama's entire presidency, the one that had surely been responsible for his long, somber walk around the South Lawn with his chief of staff. There was no use sugarcoating it. The Obamacare website was a mess.

"JUST VISIT HEALTHCARE.GOV," SAID POTUS ON OCTOBER 1, THE DAY the online marketplace opened. "There, you can compare insurance plans side by side the same way you'd shop for a plane ticket on Kayak or a TV on Amazon."

This was a bold claim, belied only by the fact that it wasn't true. On October 1, Americans could indeed type *Healthcare.gov* into their browsers. But the idea that anyone could actually *use* the site was, to put it mildly, a stretch. A lucky handful purchased coverage. For millions more, buying insurance through Obamacare was like returning a defective Kvartal at IKEA multiplied by a Comcast customer service rep. The user interface made airplane cockpits appear straightforward. The pages loaded so slowly they might as well have been written by hand.

That was assuming you could access the site at all. One reporter set out to create an account, and succeeded—after sixty-three tries.

By now, Healthcare.gov's early faults have been well chronicled. Less commonly understood is the way the website was to Obamacare what LeBron is to the Cleveland Cavaliers. With the star player sidelined, everything else began falling apart.

Nowhere were the consequences more dire than with a simple, ten-word promise: "If you like your plan, you can keep your plan." Between 2008 and 2013, President Obama made some variant of this pledge approximately three dozen times. This was surprising. Ordinarily, such blanket statements drew a flurry of angry fact-checks.

If you like your plan

Flagging that people with objectively crummy insurance plans may nonetheless like them.

. . . you can keep your plan.

Insurance companies might be forced to drop plans that don't meet the law's higher standards. Just flagging.

These caveats were ignored, and in theory shouldn't have mattered. The Americans at greatest risk of losing insurance had overpriced, subpar plans to begin with. Surely they would welcome the online marketplace with open arms. Imagine learning you could trade in your beat-up '92 Civic for a brand-new Lexus. You wouldn't be furious. You'd be thrilled.

With no Healthcare.gov, however, there was no marketplace. The Lexuses were locked inside a labyrinth of buggy code. Meanwhile, the '92 Civics were being unceremoniously scrapped. Four million Americans were informed their old insurance would soon be taken away.

"If you like your plan" was not a lie, exactly. It was an act of unfounded optimism rather than willful deceit. But when people are terrified about losing health care, such distinctions cease to matter. For the first time ever, President Obama's credibility with voters began to erode.

This is why the Obamacare rollout was not merely a hiccup or a growing pain. It was a disaster. Imagine if every flight sold through Kayak wound up, unexpectedly, in Mogadishu. Imagine if Amazon accidently began shipping nests of angry hornets to people's homes.

And here's where the business world and political world diverge. If Kayak or Amazon are in crisis, it's a story. If a presidency is in crisis, it's *the* story. In the White House, good news is fleeting, and bad news is *Groundhog Day*. Night after night, Americans heard the exact same report. President Obama's site wasn't working. His promises weren't being kept.

Then there was the most dispiriting news of all: outside the building, people had tried to warn us. In March, a team of McKinsey consultants predicted that Healthcare.gov might not be ready on schedule. In July, the site failed key tests. In August, insurance companies briefed Nancy Pelosi's office with concerns. In hindsight, I could even recall my own father intervening. "I have this friend in Connecticut," he told me that summer. "He says these sites are going to be a mess."

This wasn't entirely random. My dad happens to be a doctor. But he didn't work in government. He certainly wasn't the president of the United States. So what accounted for our massive blind spot? Put another way, how is it that Andy Litt knew Healthcare.gov was going to be a shitshow and Barack Obama did not?

While I'll never be entirely certain, here's my guess: no one told POTUS his law was in trouble for the same reason no one told him his head might be set aflame. What we needed was a designated pessimist, someone in charge of sounding the alarm. But role of the naysayer hadn't been assigned to anyone. As a result, no one spoke up.

In Obamaworld, the typical response to crisis was to circle the wagons, but Healthcare.gov was no typical crisis. Not long after the government reopened, POTUS announced that a cadre of patriotic programmers was being flown in from California. The fate of the second term—the future of hope and change—was no longer in our hands. Only Silicon Valley's disruptors could save us now.

IN DECEMBER, PRESIDENT OBAMA'S APPROVAL RATINGS FELL TO 40 percent, the same number George W. Bush hit three months after Hurricane Katrina dealt his presidency a mortal blow.

"What's the mood like over there?" friends asked, less in a spirit of concern than rubbernecking. The Obama White House was imploding. They wanted juicy details. Determined not to give them the satisfaction, I told them we were just fine.

"Everyone's feeling great!" I promised.

But I was lying. In truth, the mood was awful. We slouched through the EEOB's hallways. We plodded across West Exec. It was as if every White House staffer had been simultaneously broken up with.

And in a way, we had been broken up with. The most pressure-packed workplace in the country was no longer located at 1600 Pennsylvania Avenue. Overnight, the center of the universe had left us for a nondescript building in Maryland, the headquarters for our heroic team of nerds. I rooted for what reporters were calling "the tech surge." I wanted it to succeed. But on a more personal level, I wanted life to go back to normal. I wanted the White House to feel White House-y again.

Part of me doubted it ever would. Websites can be fixed, but the damage to Obama's movement would be far more difficult to repair. In Washington, you often hear people say that "perception is reality." I usually like to imagine these people walking through plate-glass windows. This time, however, they had a point. After our shutdown triumph, the Obamacare rollout was supposed to be the knockout punch. The battle over the size and scope of government—an argument decades in the making—would be settled once and for all.

Instead, thanks to unforced errors, our greatest victory had become our most catastrophic defeat. To the press, Obamacare would be forever mired in controversy. To the law's opponents, Healthcare.gov would be a symbol of government's inability to do anything right. It was disappointing. It was infuriating. More than anything, it was exhausting. Sure, we were on the right side of history. But would we ever actually prevail?

Perhaps it sounds corny, but what kept me going was Zoe Lihn. "I refuse to walk away from forty million people who have the chance to get health insurance for the first time," said POTUS, and despite my growing doubts, I agreed with him. I thought about Zoe's mom, Stacey, standing on that Charlotte stage. I thought about Wendy, my Ohio volunteer, pacing our campaign headquarters despite near-crippling pain.

And I thought about Jacqui. A year earlier, after earning her law degree, she had graduated into one of the toughest job markets new lawyers had ever encountered. While she found work that paid decently, it didn't come with benefits. She was uninsured.

I should add that Jacqui was in perfect health. Her situation was nothing like Wendy's or Zoe Lihn's. Yet this was precisely the point: my girlfriend was one of the lucky ones, and her lack of coverage was something we worried about literally every day. The night Jacqui first met my family, we were walking to a bar in Brooklyn when my sister accidentally brought a spike heel down on her foot. More than her pain, I remember her panic. Was something broken? How much would an emergency room cost? Could we set a bone using duct tape and WebMD?

There we were, two well-educated professionals in the richest country on earth, wondering if we knew how to jury-rig a splint.

OUR A-TEAM OF CODERS COULDN'T UNDO THE DAMAGE TO THE president's credibility. They could, however, make Healthcare.gov suck less. By mid-December, the site's error rate was under 1 percent. In speeches, we urged Americans to try signing up online.

In private, however, I advised Jacqui to wait. Engineers were still fixing faulty code. The insurance wouldn't actually kick in until New Year's. Why not give it more time?

Specifically, why not wait until Christmas? Two years into our relationship, I had finally promised to spend Jacqui's favorite holiday in New Jersey with her family. I saw this as a gesture, like telling an English friend you'll go with him to watch a soccer game. She saw it as a full-blown embrace, like telling an English friend you'll tattoo Saint George's Cross on your forehead and beat a Scotsman with a pipe.

Thus my real reason for suggesting Jacqui wait to buy insurance. If the Christmas spirit overwhelmed me, I could always count on Healthcare.gov to darken the mood. Driving north on the turnpike, we listened to "Grandma Got Run Over by a Reindeer," and I casually wondered who paid for Grandma's outpatient therapy. A nativity scene became an excuse to mention that, thanks to the Affordable Care Act, pregnancy was no longer a preexisting condition.

And yet, despite my determination to be a wet blanket, Christmas charmed me. It wasn't just the presents and pajama-clad mornings. Beneath the cheer and sleigh bells, I found a web of often-contradictory traditions. Layers of familial angst needed managing. There was the constant sense that one's celebrating was not quite celebratory enough. And to think I worried about fitting in! Pork products notwithstanding, Christmas was the most Jewish thing I'd done in months.

Lost in piles of wrapping paper and mountains of spiral-cut ham, we nearly didn't get around to health care. But once Christmas Day was over, we decided it was time to take the plunge. On the night of December 25, 2013, Jacqui and I closed the door to her childhood bedroom. Then, cautiously, we opened Healthcare.gov.

When speechwriters sought out Real People for POTUS remarks, we tended toward the dramatic: hero firefighters, single moms returning to college, wounded veterans beating the odds to recover. If an anecdote wasn't extraordinary (or, as often happened with midwesterners, the RP was frustratingly humble), I wasn't above teasing out a quote.

"So, would you say getting hired at the Chrysler plant made you believe that, in America, anything is possible?"

"Uh, I guess so."

"No, really. Would you say it?"

What I had forgotten, until I myself became a real person, were the countless far less dramatic stories that begin with a decision a politician makes. Red light cameras are installed to compensate for a budget shortfall, and suddenly you're opening a ticket in the mail. A trade deal stalls in Washington, and a dentist in Malaysia buys a Hyundai instead of a Ford. A president makes health care reform a priority, and five years later a pair of twenty-somethings sit on a twin bed, wondering if their relationship will survive history's most infamous online store.

The first step, creating an account, went smoothly. But if signing in was easier than expected, what followed was hopelessly complex. To help you pick the right insurance plan, the site asked a series of questions. These reminded me of the "would you rather" game I used to play on the middle school bus. Would you rather fart every time you blink or have hiccups forever? Have no nose at all, or a second nose on your butt?

For a healthy young person, comparing health care plans involves choices no less absurd. Would you rather pay a little less if you get sick in your home state, or a lot more if you get sick anywhere else? Spend more nights in the hospital you don't need to go to, or save money treating the cancer you don't have? After an hour, Jacqui had made no progress whatsoever. Exhausted, we retreated to the kitchen to eat leftover ham and give her Republican father a chance to gloat.

Then we returned to battle. After thirty more minutes, Jacqui finished answering questions. But the moment "would you rather" was over, the glitches began. We attempted to compare insurance plans, only to discover that the least-expensive options—the bronze level—didn't seem to exist. We reloaded the site over and over. They wouldn't appear. There are only so many times you can refresh the same page before you begin to wonder if you're in some cruel psych experiment. But with no better options, we forged ahead. Finally, after approximately ten million refreshes, our persistence was rewarded. The full list of health care plans appeared. Jacqui chose one. The site asked her to wait.

So we did. We waited and waited and waited and waited, watching a little glowing circle chase its tail in the middle of the screen. Finally, just as we were about to give up hope, something began loading. A new screen appeared.

It was the home page. Every bit of information was gone. We would have to start from scratch.

The fight that followed was quiet. We didn't want to give Jacqui's dad the satisfaction of knowing Obamacare was tearing us apart. But we made up for volume with intensity, alternating between personal and political in a whiplash-inducing weave.

"You told me the site was working . . ." whispered Jacqui.

"Well, you don't give the tech surge enough credit."

"They don't deserve it!"

"You overreact about everything!"

"You're not really supportive!"

"You always blame me instead of Congress!"

"Oh yeah? Well, Democrats don't care about the middle class!"

Long-drawn-out silence.

"You. Take. That. Back."

After a while we stopped arguing. What choice did we have but to log back in? Jacqui created her new account. She reanswered her questions. She repicked her plan. Once again, we waited for what felt like forever, holding our breath as the glowing circle spun. Finally, something began loading. A new screen appeared.

It was a confirmation page. Jacqui was insured.

I won't say that everything became perfect in that moment. All I will say is that we jumped off that ancient twin bed and hugged each other more fiercely than we ever had before. We were crying, not with frustration this time, but with joy. True, Obamaworld had mishandled its most cherished legislative priority. True, Republicans in Congress were gleefully watching our approvals slide. But in that tiny room in New Jersey, those things didn't matter.

Here's what did matter. Barack Obama fought to make insurance affordable for everyone, well past the point where it made political sense. He made mistakes. He had blind spots. Sometimes, he even let us down. But he never gave up. He never walked away from Jacqui. And because he stood by her, the person I loved would now be able to see a doctor if she got sick.

As far as I was concerned, it was the most wonderful time of the year.

12

IN THE BARREL

And then came 2014, which just sucked. Top to bottom awful. The worst.

Consider the following encounter. Early that spring, while walking home from work, I noticed a short, plump woman approaching me in a kind of determined shuffle. I guessed she was in her late forties or early fifties. Her hair was gray and spiky, and she wore a sweater with both the color and texture of whole-grain bread.

"Excuse me," she asked. "Did you work for the Presidential Inaugural Committee?"

It's not uncommon for certain White House staffers to be recognized. One morning in 2009, when I was still at the Crisis Hut, my fellow intern Sonia floated into her cubicle on a cloud.

"I just saw Jon Favreau at the Whole Foods!" she exclaimed.

That was her entire story. She hadn't spoken to the president's chief speechwriter. She wasn't carrying his child. She merely glimpsed him browsing overpriced lettuce. Apparently, that was enough to rock her world.

But for every staffer who becomes a local celebrity, at least a hundred remain obscure. For my shyer colleagues, anonymity was the best part of public service. For the rest of us it was a trade-off, no different than writing songs for Rihanna or designing Dwyane Wade's line of shoes. At least, that's what I had always told myself. Now, standing on the sidewalk and basking in the glow of a stranger's attention, an itch was being scratched. *I've been noticed! I have a fan!* I smiled so broadly my lips hurt.

"That's right!" I announced. "I spent six weeks on the Inaugural Committee." My spiky-haired admirer nodded, thrilled to have her suspicions confirmed.

"And didn't you also work for Tim Kaine?"

Now this was really stunning. Imagine meeting Joseph Gordon-Levitt and bringing up his Pop-Tarts commercial from 1991. Mentally, I upgraded the woman to superfan. Then I addressed her with the perfect blend of modesty and poise.

"I helped with Senator Kaine's 2012 convention speech," I confessed. "But it was very good to begin with. I barely had to make changes." Wowed by my down-to-earth demeanor, she nodded even more eagerly than before.

"And now you work in the White House, right?"

At last! Here was the question I had been anticipating.

"I do indeed," I said, dropping even the pretense of humility. "Actually, I started writing speeches for the president when I was just twenty-four years old."

The woman's eyes grew large. I basked preemptively in her praise. Then, without warning, she jabbed a finger in my face.

"I know you!" she cried. "You stole my inauguration tickets!"

"Wh . . . ?" I stammered. But before I could get a word out, she escalated the charge.

"You stole my tickets!" she shouted. "You're a racist! You're racist, you're a criminal, you're in the KKK!"

Ordinarily, I would have rushed to defend myself. But this woman was such an expert on my life story. For a moment, I actually wondered if she knew something I did not.

"I . . . I don't think so?" I suggested. This did nothing to improve matters. When my accuser spoke again she was even louder, this time for the benefit of passersby.

"He's racist! He's criminal! He's KKK!"

I briefly considered reasoning with her. A heart-to-heart chat, a little active listening, and surely we'd be back on the same page. Then the spiky-haired woman resumed yelling, and I revised my view and fled. Half walking, half running, I raced down the sidewalk in my dress shoes, my superfan shuffling behind me in pursuit.

THAT, IN A NUTSHELL, WAS 2014. WE DID OUR JOBS. WE WERE PROUD of our accomplishments. Everyone hated us anyway.

Our catalog of woe began, as it always seemed to, with Obamacare. (If you find yourself wishing we could move past health care already, just imagine how we felt.) As the New Year dawned, reports weren't all bad. People were finally using Healthcare.gov. But an insurance market is like the punch at a fraternity party, where young, healthy people are the mixers and old, sick people are the booze. Get the proportions right, and the whole thing goes down smoothly. Fail to mask the alcohol, and you're screwed.

At the beginning of 2014, we were looking at a ten-gallon bucket of Everclear. Young people were tuning out information about Obamacare. They didn't know Healthcare.gov was finally working, or that in most cases they could purchase insurance for less than a hundred bucks a month. The first window to buy coverage closed in March. If we couldn't grab the attention of

America's youth by then, Obamacare premiums would skyrocket for everyone else. That would trigger what is known, in otherwise sedate policy circles, as a *death spiral.* The White House communications department was ready to throw a Hail Mary pass.

What we didn't know was that the perfect play had been drawn up five years earlier. On Halloween 2008, a producer for the comedy website Funny Or Die had been seized with patriotic fervor. Grabbing a pen and a piece of unlined paper, he began to write.

> *I, Mike Farah, guarantee that the FOD team will for sure meet or have the opportunity to meet Barack Obama between October 31st 2008 and October 31st 2016 or I'll eat my hat.*

It is safe to say that, at the time, hat eating was the likely outcome. But Farah knew how to put himself in the path of lightning. In early 2013, he was invited to his first White House meeting. For the next twelve months he hovered in the background, never pushy but always available, like an extremely well-tanned Secret Service agent. His company produced *Between Two Ferns,* a weird online talk show hosted by comedian Zach Galifianakis. Maybe POTUS could appear as a guest? The idea was ludicrous. And then it wasn't. In 2014, when desperate times called for desperate measures, Farah was there to help.

Here I must confess something: I thought putting POTUS on *Between Two Ferns* would be a huge mistake. My concern was not that he would come across as unpresidential. My concern was that he would come across as kind of a dick. The camera didn't always catch the nuance in his humor, the way a smile or raised eyebrow could soften a rhetorical blow. Good-natured teasing in person could look like bullying on-screen.

Take what happened when President Obama met my parents. This was backstage before a speech at a New York City Sheraton, the kind of photo line POTUS had done countless times before.

"Mr. President," I announced when my turn came, "this is my mother, father, and sister."

"Mom, Dad, sis, good to see you!" (To remove the potential for hurt feelings, the president avoided "Nice to meet you" at all costs.)

We grouped ourselves into a bell curve by height. POTUS put his arm around my shoulder, and I readied myself for our click. But instead of a camera, I was surprised to hear President Obama's voice.

"Uh, maybe you should put down your speech?"

My printed draft! I had taken a copy off the plane, and now, to my horror, I realized I was clutching it like a security blanket. I trotted over to a nearby aide, handed him my file folder, and sheepishly slouched back into the frame. POTUS once again put his arm around my shoulder. But just as quickly, he released.

"Your badge?"

This was less playful. It was the tone Jacqui used when I was about to forget my credit card at a bar. "You're still wearing your badge," he repeated. "You might want to take that off."

The president was still smiling. But this was no longer his bright, photo-line smile. It was the can-you-believe-this-idiot smile he reserved for staff who did something harmless but dumb. As fast as I could, I yanked my ID off my neck and I stuffed it forcefully into my pocket. Naturally, this caused the lanyard to spring back out. I wrestled the lanyard in. It popped back out. I wrestled it in. It popped back out.

Finally, after what felt like an eternity of struggle, I secured my badge. I was ready for our picture. But there was no way

POTUS was letting me off the hook. Instead, he turned toward my father.

"You know," he said, grinning slightly, "he's a pretty good writer. But he's a little absentminded."

How did President Obama do that? Did my dad write a letter to the White House, suggesting the single most accurate adjective with which to embarrass his son?

I wouldn't put it past him—unlike me, he wouldn't have forgotten to attach postage. More likely, however, the president's extraordinary ability to read a policy memo extended to reading people as well. Just as he could pluck the most important issue from pages full of jargon, he could scan a human being and intuit their most cringeworthy trait. He wasn't being mean, exactly. The teasing was always in good fun. But behind his trash talk inevitably lay a kernel of deeply personal truth.

That's what worried me about *Ferns*. If I'd been watching that exchange with my parents on YouTube, rather than living through it, there's a good chance POTUS might have appeared cruel. How could he trade barbs with Zach Galifianakis, one of the best-loved schlubs in America, without going too far?

The discomfort ran the other direction as well. Rachel Goldenberg, one of the episode's producers, later told me no one knew if the host truly had permission to read his lines. At one point, he struggled.

"What does it feel like . . ." He hesitated, but POTUS was having none of it. "Come on, man!"

With the president's permission, Zach tried again.

"What does it feel like to be the last black president?"

I wish I could take credit for jokes like that. But when we worked with professional comedians (and Zach, Scott Aukerman, and B. J. Porter, *Ferns'* creators, are three of the best) I tried not to butt in. Instead, I added a few words to the president's plug for Obamacare. I made sure he mentioned a phone number, just in case Healthcare.gov unexpectedly crashed. Mostly I waited and worried, certain our last-gasp attempt at pitching health insurance to young people was doomed to fail.

Here's how wrong I was. In July 2013, I drafted a health care speech for POTUS. As I write this, it has about ten thousand YouTube views. In the twenty-four hours after the release of *Between Two Ferns,* eleven *million* people watched it online. Put another way, for every dining room table you could fill with people who have streamed my speech, ever, you could fill Radio City Music Hall with people who saw *Ferns* in a single day.

And they didn't just watch. They took action. Overnight, traffic to Healthcare.gov jumped 40 percent. In the two weeks that followed the episode's March 11 release, the young and healthy came out of the woodwork to sign up.

Back in 2013, before Healthcare.gov, the nonpartisan Congressional Budget Office predicted seven million people would purchase insurance through Obamacare. After the site's disastrous launch, the law's critics confidently gloated that this projection would be way off. They were right. By mid-April, eight million people had signed up.

On April 17, President Obama held a press conference to trumpet the good news, and Cody assigned me to write what's known as the topper—brief, prepared remarks before the first question is asked. I was thrilled. I couldn't wait to help POTUS tell our comeback story.

Yet when I arrived outside the Oval Office a few minutes before the presser, the president didn't appear excited. In fact, he looked downright glum. As usual before a public appearance, he was bantering, this time with press secretary Jay Carney. But the bouncy energy he displayed before big rallies was missing. At one point, the subject turned to weekend plans.

"Golfing again?" Jay joked. GOP frenzy over the president's weekly eighteen holes renewed themselves like clockwork each spring, a blooming crocus of outrage. But President Obama didn't look amused. He looked weary.

"Jay," he said, "that's the only time I get to go outside."

It was only once we arrived in the briefing room that I understood why POTUS might feel deflated. On television, presidents at press conferences appear larger than life. In person they look like goldfish in a bowl. The James S. Brady Press Briefing Room—which holds forty-nine seated reporters—is only slightly bigger than your average American garage. The stage has all the height and majesty of a shipping pallet. President Obama wasn't commanding his audience. He was surrounded by it.

Before I continue, I should clarify something: I don't think most members of the traditional White House press corps are biased toward a political party. That said, they are definitely out to get the president. It's only natural. Every reporter wants to be Woodward and Bernstein. Setting their sights on anyone but POTUS would be like Captain Ahab chasing a guppy across the high seas.

Along with this common dream—uncovering the next Watergate—reporters share a tendency to herd like wildebeests around a single dramatic arc. "What's the story here?" they ask, as though narratives are distant planets to be discovered rather than frameworks they create. As a consumer of news, I empathize. I count on journalists to help me decide what matters and what does not. But the danger arises when facts start to fit the narrative and not the other way around. Combine reporters' desire to congeal around a story line with their innate distrust of the powerful, and a presidential comeback is hard to stage.

Still, I was confident that our Obamacare news could pull it off. Imagine if, after its brush with the iceberg, the *Titanic* had not just stayed afloat but crossed the Atlantic ahead of schedule. That's what had just happened to POTUS's top domestic priority. Surely we could all agree this was a game changer.

Well, no. We couldn't. The eight million new signups, touted so enthusiastically in the president's topper, were almost completely ignored once the reporters' questions began.

When will your health care law become popular?

Can you finally fix what's wrong with it?

Will Democrats embrace it on the campaign trail?

These were not unreasonable things to ask. What was unreasonable was the set of assumptions behind them. With the early failure of Healthcare.gov, the narrative had reached its "floundering president" chapter. No amount of evidence could turn the page.

PRESIDENT OBAMA HAD A TERM FOR BEING STUCK IN THESE CYCLES of negative news coverage. He called it being "in the barrel," and he treated it the way kindergarten teachers treat an epidemic of head lice. Deeply unpleasant. Happens once or twice a year. Eventually goes away.

I was less certain. I still worried that Americans saw the Healthcare.gov debacle as a kind of infidelity, one they had not yet decided to forgive.

As the 2014 Correspondents' Dinner approached, I didn't think one night of comedy could fix everything. That said, I thought it could help. If POTUS was ambitious in his material and flawless in his delivery, voters might feel the old spark rekindled. Reporters might decide the narrative wasn't so set after all. As the joke-writing process began, our task seemed clear. Take big swings.

And not just at our opponents. By making fun of our own low moments, we could show the world we had moved on. This concept—self-deprecation—is one that a surprising number of important people fail to grasp. I know far too many speechwriters who have lived through some version of the following exchange.

> POLITICIAN *(doing his best cool dad impression):* I love making fun of myself! Whadaya got?
>
> SPEECHWRITER *(nervous):* Well, I was thinking we could joke about the idea that you're kind of a diva?
>
> POLITICIAN *(recipient of a sudden personality transplant):* What? A diva? Why would anyone find that funny?

If you ever find yourself on the receiving end of this question, here is my advice: Do not answer it! Fake a seizure. Play dead. Flee the country. Whatever you do, don't open your mouth.

Where self-deprecation was concerned, President Obama's joke writers were lucky. While I can't say he truly enjoyed making fun of himself, he understood its value. No less important, he had a sense of gallows humor. A friend of mine who worked in the National Economic Council was first introduced to POTUS at the end of 2013.

"This is David Edelman," said my friend's boss. "He's working on tech issues."

Without missing a beat, POTUS eyed him skeptically.

"You're not the guy who designed my website, are you?"

There was another reason the president liked making fun of himself: it earned him the right to mock people who genuinely pissed him off. In our first Oval Office meeting for the 2014 dinner, our draft included a full page of one-liners about Healthcare.gov. In exchange, we got to make fun of everyone from Mitch McConnell to the right-wing billionaire Koch brothers to the hosts of *Fox & Friends*.

We also took a shot at the growing number of Republicans obsessed with Vladimir Putin. This was a truly bizarre trend.

Prominent conservatives had recently begun heaping praise upon the Russian autocrat, often in not-quite-unsexual terms. "I know the only time that Vladimir Putin shivers is when he takes his shirt off in a cold Russian winter," Governor Mike Huckabee declared, as if he'd read a particularly ham-fisted profile on Grindr. Sean Hannity and Rudy Giuliani were less swoony, but not by much. With the help of our graphics department, we came up with an image to meet the moment: Mike, Sean, and Rudy at a slumber party, giggling over a poster of a shirtless Vlad.

We added the slide to our growing list of audiovisual bells and whistles: more than a dozen photoshopped images; a parody of *The Matrix;* a video starring Julia Louis-Dreyfus as her character from *Veep*. We even held a brief photo shoot where the First Lady, looking deeply confounded, held up a picture frame made of popsicle sticks. Then we added the resulting picture to the slide deck. Tech-wise, it was our most ambitious Correspondents' Dinner to date. No one could accuse us of not swinging big.

"You're sure all this stuff will work?" asked Cody. "After health care, we can't have any screw-ups."

"I've got this," I promised. "We'll do a run-through the morning of the speech."

THE RUN-THROUGH IN QUESTION TOOK PLACE ABOVE THE HILTON ballroom, in a small, enclosed catwalk about thirty feet above the stage. The floor was made of concrete. Wires splayed in all directions. At the far end sat a gloomy metal cage that brought to mind a third-world zoo.

Inside the cage sat Steve. Steve did not work for the White House. Steve worked for the Washington Hilton. From the moment we met, he made it clear he considered the Hilton the more important institution of the two. With his close-cropped hair, permanently put-upon expression, and belly perfectly designed for resting cups of coffee, Steve oversaw the A/V booth with the territorial instinct of a panther and the work ethic of a house cat.

The only other person at the run-through was Jenn, the newest member of the White House graphics team. The kind of person who wore purple fanny packs without irony, Jenn also served as fan club president for the Canadian rock band Rush. She was not the type to keep her emotions in check.

And right now she was nervous. "Does everything look okay? Do you need me to bring a backup computer? Is there anything we need to triple-check?" But Jenn's anxiety only made me calmer. The cool, battle-hardened veteran, I promised her everything would be fine.

My sense of icy, been-there-done-that composure was further reinforced when I returned to the Hilton that evening. There, a CAT team member checked the pin on my tuxedo jacket, lowered his assault rifle, and ushered me into the hotel's back entrance with a nod. I stalked the predinner receptions for finger food. I made the president's last-minute edits. Then, as POTUS prepared to address the crowd, I headed for the catwalk.

Jenn was already there, MacBook at the ready, hyperventilating in a flowing purple gown. Steve was there as well. His expression suggested he had DVR'd something—*Ice Road Truckers,* maybe—and held us personally responsible for keeping him from his show.

As the president's monologue began, however, I was too high on adrenaline to give either of them much notice. This was our chance to get out of the barrel, and with a jolt of excitement, I realized it just might work. The *Veep* video was a hit. The self-deprecating section on health care left the crowd impressed and applauding. And POTUS's timing—already good to begin with—grew better every year. In the ballroom below, he landed joke after joke. Before I knew it, the only thing standing between us and a huge success were a few remaining slides.

"Two weeks ago, Senator Ted Cruz and I, we got a bill done together."

"Okay," I said from the catwalk, "go to slide."

This was where Steve came in. Stretching out a reluctant arm, our A/V expert pushed a large square button. On the giant screens in the ballroom, the live feed of POTUS was replaced by an image of hell freezing over. The audience laughed.

"Okay, go back," I said. Before releasing his button, Steve was supposed to wait for my cue. But he had already resumed splaying grumpily in his chair. The slide had vanished from the screen.

POTUS continued his monologue, unaware that above him, a passive-aggressive war was breaking out. With each slide—Biden shoe ambush, *Game of Thrones* staff meeting, Raging Socialist High School—Steve took just a little bit longer to press his button. Each time, he gave a slightly more exasperated scowl before slumping back into his seat.

"Everywhere I look, there are reminders that I only hold this job temporarily," said POTUS.

"Okay, go to slide."

Steve once again leaned forward. But this time, too busy loathing me to pay attention, he failed to place his finger in the center of the button. Instead he clipped the side. To the confusion of the audience, a picture flashed on-screen, then disappeared. I was furious. Steve was jeopardizing President Obama's big moment! This was completely unacceptable!

And that's when I noticed that our next slide was missing entirely.

In a state of shock, I double-checked Jenn's laptop. There was Slide 13—the president, looking frowny, standing in the Oval Office. And there was Slide 15—computer code from *The Matrix*. But Slide 14, the picture of the First Lady holding her popsicle-stick picture frame, was nowhere to be found. Slide

13; Slide 15. Frowny-face Obama; *The Matrix*. There was nothing in between them. And POTUS had no idea. In less than ten seconds, the president of the United States would be humiliated, and it would be my fault.

By now, Jenn realized what was happening and, leaping into action, she panicked. "Where's the slide? Where's the slide?" she gasped, in a tone I thought was reserved for missing children. "There is no slide," I replied, in a tone I thought was reserved for climbing the guillotine steps.

Steve, meanwhile, was in heaven. This was way better than *Ice Road Truckers*. As the seconds ticked away in slow motion, he rested his hands contentedly on his belly, as though we had all learned a valuable lesson about not bothering the A/V guy. Jenn had a series of small strokes. I stared at the tiny monitor in the catwalk, totally numb.

POTUS, meanwhile, kept going. "George W. Bush took up painting after leaving office, which inspired me to take up my own artistic side."

The president paused, waiting for an image only three people in America knew did not exist.

"I'm sure we've got a shot of this," the president said. My skin had turned clammy. My mouth was sandpaper dry. POTUS licked his lips, annoyed.

"Maybe not."

I have never been the kind of person who refers to "my career" as if it is something I gave birth to. Still, it was distressing to watch it die. President Obama was about to look foolish. A new wave of "hapless president" stories would soon dominate the news. Heads would roll, beginning, understandably, with mine.

But then, just as I was wondering if Silicon Valley recruits disgraced former speechwriters, something happened. In the blink of an eye, POTUS composed himself. He looked out across tables full of reporters, each one eager to write about yet another

White House screw-up. Then he broke into an exasperated grin. "The joke doesn't work without the slide," he explained. Suddenly, the audience was in on the secret. They began to laugh.

"Oh well." Another pause. "Assume that it was funny." He chuckled at his own ad-lib, and the audience joined him. Then he turned to Joel McHale, the night's headliner.

"Does this happen to you, Joel?"

It was not the highlight of the evening. But neither was it the story of the night. POTUS finished his ad-lib, delivered his serious close, and took his seat. The crowd stood to applaud. I sat in Steve's cage inside the catwalk, grateful for the president's quick thinking.

Still, another chance to escape the barrel had been squandered. And changing the narrative was about to become harder than ever. Summer was almost upon us.

NEVER WORN A SUIT AND TIE TO WORK DURING A D.C. SUMMER? Want to experience it for yourself? It's easy! First, wrap something snug around your neck, like a scarf or boa constrictor. Next, drape yourself in thick, barely breathable material. Wool works nicely. So does a hefty coat of tar. To approximate the discomfort that results from hot sun and black leather dress shoes, wrap your feet snugly in a colony of fire ants. Last but not least, fill a stockpot with water. After heating to a gentle simmer, climb inside it. Now try to do your job.

Women, who could wear skirts or dresses, had it easier during summers. But not by much. Whether you work in a T-shirt or full military uniform, Washingtonians are united by the season-long schvitz they take each year. It builds character. It's a good subject for small talk. It is not conducive to running a country. Brains slow-cook in their owners' skulls. Adults who can afford long vacations flee the district, leaving twenty-three-year-olds disproportionately in charge.

Perhaps this is why summer was to Obamaworld what winter is to *Game of Thrones*. Madness flourished like mosquitoes in the heat.

August 29, 2008: Sarah Palin is nominated for vice president.
August 7, 2009: The term *death panel* is coined.
August 28, 2010: Glenn Beck holds his "Restoring Honor" rally on the steps of the Lincoln Memorial.
July 30, 2011: The single worst day of the debt ceiling crisis.
August 30, 2012: Clint Eastwood speaks at the Republican Convention. (This one backfired, but it was such a bad idea that it still counts.)
August 1–30, 2013: Government shutdown fever takes hold.

Seriously, what was wrong with summer? And where pandemonium and general awfulness were concerned, the summer of 2014 would dwarf them all.

First came ebola, a disease from the opening scene of a Michael Crichton novel. Horrifying symptoms. Rapid transmission. No known cure. When an outbreak hit West Africa in December 2013, most Americans barely noticed. But on August 2 of the following year, a U.S. citizen was infected. Suddenly, everyone lost their minds.

Myself included. I tried my best to keep things in perspective: the American who contracted the virus was a missionary overseas; our public health system was light-years ahead of Liberia's or Sierra Leone's. But the facts failed to reassure me. Sitting on our gray Martha Stewart Living sofa, Jacqui and I planned for our new lives.

"We can camp out in the woods. I'll bring my fishing gear."

"You never catch anything."

"Okay, fair point. But maybe if we were starving I'd improve?"

Then there was ISIS, the ebola of terrorist organizations. Murdering scores of innocent people, burning victims alive, forcing women into sexual slavery: their atrocities were impossible to fully catalog, let alone comprehend. On August 7, with the group gaining territory, President Obama authorized air strikes against ISIS fighters in Iraq and Syria. On August 19, in retaliation, they beheaded an American journalist named James Foley on film.

This was a turning point. Practically overnight, ISIS went from interchangeable foreign bad guy to nationwide bogeyman. The president's critics were quick to pounce. Country Clubbers said he was soft on terror. Flat Earthers and Holy Warriors went further, accusing him of being on the terrorists' side.

The second charge was extravagantly false, and the first wasn't much better. Rather than backing down in the fight against ISIS, POTUS stepped up raids and air strikes, taking out top commanders along with scores of troops. With our military's help, Iraqi and Kurdish forces began rolling back the borders of the so-called Islamic State. But if fearful Americans were looking for a leader of the chest-thumping, why-I-oughta-kill-those-dang-terrorists-myself variety, they had voted for the wrong guy. Earlier that year, when POTUS and his foreign-policy team were asked to define an "Obama Doctrine," they settled on the following:

"Don't do stupid shit."

Inspiring, I know. But what the slogan lacked in soaring vision, it made up for in common sense. President Obama wasn't opposed, on principle, to throwing America's weight around. He simply believed the best course of action often involved self-restraint. Put another way, there are few things more dangerous than a president doing a bad John Wayne impression. He refused to indulge in clash-of-civilizations rhetoric favored by both right-wing pundits at home and ISIS propagandists abroad.

This was almost certainly the right decision. It deprived some of the world's most sadistic killers of an easy recruiting tool. But at times it put POTUS at odds with the national mood. When the Foley beheading video was released, President Obama was in Martha's Vineyard; he chose not to cut his vacation short. Also, he referred to ISIS as "ISIL." While technically accurate, this could come across as tone deaf, like a waiter who explains that it's wagyu beef, not Kobe, on which you are currently choking.

Even our political allies could be frustrated by President Obama's near-pathological calmness. Still, his message to the public was simple, never directly stated but always easy to infer. Don't freak out.

It fell on deaf ears. America was in a freak-out state of mind. And politics only made things worse. Hoping to jump-start his 2016 presidential campaign, Texas governor Rick Perry announced that ISIS might have already crossed America's southern border. Not to be outdone, fellow candidate Rand Paul insisted you could catch ebola by standing near the wrong guest at a cocktail party. Cable news networks occasionally pointed out that these claims were nonsense. More often, they luxuriated in the kind of ratings boost only a national panic could provide. One day I happened to glance at MSNBC. There, live on camera, the Reverend Al Sharpton was being taught to don a hazmat suit.

Uh-oh, I thought. *This can't end well.*

I was right to be worried. The summer of 2014 wasn't the first time terrible things happened. But it was the first time I began regularly hearing the phrase "The world is falling apart." Objectively, this just wasn't true. In 2014, terrorists killed eighteen people in the United States; approximately the same number were killed by cows. Your odds of dying a violent death, being assaulted in a United States city, developing terminal cancer: by historical standards, all these were near their all-time lows. Yet the number of Americans certain the sky was falling rocketed to new heights.

I blame the media. More accurately, I blame social media. Thanks to Twitter, you could follow every tragedy as it unfolded. Thanks to YouTube and Facebook, you could now watch, rather than merely read about, the world's most horrific events. We were experiencing less violence than almost any generation in human history. But we were witnessing more violence than ever before.

There is, I must admit, another reason I may have been less frantic about the universe than my peers. At the very moment civilization appeared to be collapsing, my speechwriting career was really taking off.

AT THE WHITE HOUSE, THE UPPER ECHELON OF STAFF IS MADE UP OF "commissioned officers." Commissioned officers, in turn, are grouped into three tiers. Special Assistants to the President are outranked by Deputy Assistants to the President. Deputy Assistants to the President are outranked by Assistants to the President. Assistants to the President, in theory at least, answer directly to the commander in chief.

Commissions cannot be handed out like candy. The total number of SAPs, DAPs, and APs (pronounced "saps, daps, and apps") is tightly controlled. So when Kyle O'Connor, a POTUS speechwriter since the first campaign, announced he was leaving his SAP spot for a job at a tech company, it was a big deal. Subtly at first, and then with increasing shamelessness, I pestered Cody for a promotion. Late that summer, he decided I had earned it. There were still dozens of DAPs and APs who outranked me. But as a newly minted Special Assistant to the President, I was now technically a member of the White House senior staff.

My new title came with a raise. Preacher Man now owed me an extra cent a year. But inside the building, where everyone was underpaid relative to the private sector, what mattered more than money was access.

In that regard, becoming a SAP was like winning a sweepstakes. A few days after my promotion became official, I made my way to the operations office, where a young woman walked me through a grab bag of new perks. I could now park inside the White House campus, order frozen yogurt from the West Wing takeout window, and reserve a table at the Navy Mess. I received a commissioned-officer certificate for framing. My business cards were upgraded with a fancy raised seal. Technically, people were even supposed to refer to me as "The Honorable David Litt," although sadly no one ever did.

The most extraordinary new benefit was the one saved for last. Opening a small folder, the young woman from operations retrieved a silver key, about an inch long and a quarter-inch thick. I recognized it instantly. It was the same key Terry wore around his neck, the key I was certain granted access to the bunker and escape pod. Now, it was being placed in my outstretched hand.

The operations associate noticed my startled expression. She met my gaze with a solemn, weighty stare.

"This," she told me gravely, "will get you into the senior staff gym."

13

BUCKET

So I had done it! I had earned the right to puff away on the elliptical next to some of the world's most powerful people. I had not, however, joined their ranks. The line between senior staff and *really* senior staff was drawn anew each morning. If you held a standing invitation to Denis McDonough's 7:30 A.M. meeting, you were one of the few people who truly had the president's ear.

If, like me, you were invited to the 9 A.M. in the Roosevelt Room, you got to participate in White House show-and-tell. In fairness, these meetings were often quite informative. Puerto Rico's debt service. Community college dropout rates. The changing nature of retirement plans. It's nearly impossible to find a subject about which someone, somewhere in the federal government is not currently geeking out. One memorable morning, a scientist spent ten minutes lecturing us on America's unique blend of topsoil. I mean this sincerely: I have never felt so patriotic about dirt.

Not every 9 A.M. was successful. There was the time, for instance, that a National Security Council director briefed us on the chaos facing villages along the Pakistan-Afghanistan border. The moment the presenter finished, a hand shot up. It belonged to one of our growing number of Silicon Valley transplants.

"I was just thinking," sighed the technologist. "If only we could teach those girls to code."

Even now, years later, I don't know what to make of that moment. On one hand, what the fuck? On the other, the Obama White House was shaped, and often improved, by these minor culture clashes. Palo Alto vs. the State Department. CEO types vs. protest-march types. Those who liked college basketball vs. those who only pretended to like college basketball. (In Obamaworld, there was no other choice.)

Then there was the culture clash that dwarfed all others. Were you the type of person who preferred to stay in Washington, creating and implementing new policy? Or did you long to hit the campaign trail?

Out of necessity, plenty of staffers kept a foot in both worlds. But just as there are chocolate people and vanilla people, nearly everyone swore allegiance to one side. Campaign people saw policy people as eggheads without street smarts. Policy people saw campaign people as simpletons without attention spans. Both sides had a point.

That said, I was a campaign person. I loved the collective intake of breath at a rally the moment a candidate took the stage. I loved watching crowds roar over words like *freedom* and *citizenship*.

Most of all, I loved winning. And not just for the speedball of adrenaline and dopamine it produced. I loved winning because, in a polarized democracy, the quickest route to change was replacing Republicans with Democrats; the greatest obstacle to change was vice versa. If day-to-day governing was like choosing the right words, elections were like choosing a language.

And it was the upcoming elections, more than the fancy new business card or even the parking pass, that made me excited to become a SAP. As a commissioned officer, I was now exempt from the legal division between official and political activity. Unburdened by the rule that stymied me in 2012, I could finally write speeches for campaigns. And not a moment too soon. With the 2014 midterms approaching, I figured POTUS would spend the fall rallying voters on his fellow Democrats' behalf.

What I hadn't realized was just how unpopular we had become. As November drew nearer, the national outlook went from dark to darker. The president's approvals steadily declined. For Democrats running for senator or governor, it was as though Barack Obama had come down with chicken pox. Barely concealing their panic, they declined his offer to join them on the stump.

There were a few exceptions. On October 28, we flew to Wisconsin to support our candidate for governor, Mary Burke. It was a freezing Tuesday evening, in a state the president had already visited countless times. Even so, thirty-five hundred people showed up. The president reminded the packed high school gym of the progress we had made together. He emphasized the upcoming election's stakes.

"Cynicism is a choice," he cried. "Hope is a better choice."

The audience erupted into rapturous applause. In that room, Obama's approval rating was a billion gazillion percent. That's another thing I loved about campaigns. In their final days, you always felt on the cusp of victory, whether you were going to win or lose.

We were going to lose. Badly. By eye-popping margins, voters chose cynicism—unless you're old enough to have seen

Casablanca when it first opened, 2014 was the lowest-turnout election of your life. Not everyone stayed home, of course. Plenty of Republicans showed up. John Boehner added thirteen seats to his House majority. Mitch McConnell needed six seats to take the Senate and picked up nine. Forget poor Mary Burke in Wisconsin. We even lost governors' races in deep-blue Massachusetts and Illinois.

The 9 A.M meeting the following morning was not one of our cheeriest. Judging from people's expressions, you would have thought there was an open casket in the room. Policy people were grieving because the outcome made no sense. How could we rescue an economy, take out bin Laden, lay out proposals that routinely garnered support from a majority of Americans, and still get utterly whomped? Campaign people understood that logic doesn't decide elections. But this same understanding had allowed us to hope, however illogically, that the midterms might swing our way. It was hard to tell who was more distraught.

It fell to David Simas, the president's political director, to analyze our defeat. Unlike the rest of us, Simas held local office before coming to Washington. He still had politician hair to prove it—a well-defined part just right of center, with two brown cliffs on either side. But the face beneath them, ordinarily home to a cautiously optimistic smile, looked haggard. Each of his PowerPoint slides was bleaker than the last. Too many independents were sick of us. Too many Democrats were tired of us. This wasn't as bad as 2010. It was worse.

Then, halfway through Simas's morbid presentation, the door directly to his right burst open. President Obama walked in.

Instantly, the mood transformed. Dispirited eyes lit up. The applause echoing through the Roosevelt Room rivaled anything on the campaign. Once the cheering finally faded, POTUS began to speak.

"I hate losing," he said, "and we lost bad last night. But I intend to squeeze every ounce of juice out of these next two years."

For several minutes, he continued in this vein of dogged optimism. I remember trying to pay attention to what he said. But I couldn't. I was too busy watching the audience. Seated around the giant wooden conference table were the *really* senior staff, eyes full of misty, overwhelming gratitude. It was as though they were young children, and Barack Obama was the older brother they adored.

I knew that look. I knew how it felt to trust Obama unreservedly, unblinkingly, no matter how daunting the odds. And I realized, with a mix of horror and heartbreak, that I didn't feel that way anymore. Yes, we had cleaned up George W. Bush's mess. But that wasn't the vision that put me and so many others under Obama's spell. The promise that transfixed us was the one delivered on that January night in Iowa:

"Faced with impossible odds, people who love this country can change it."

Now it was clearer than ever. The speech that changed my life was just another act of unfounded optimism. No matter how well chosen the president's words, our story would be written in Mitch McConnell's language. The cynics were going to win. The believers were going to lose.

The pep talk ended and POTUS left. I clapped and smiled. What else could I do? But how foolish that seemed now. How naive.

AT MOMENTS LIKE THESE, I SINCERELY ENVIED BO, THE OBAMAS' Portuguese water dog. Bo did not care about poll numbers or stalled legislation. Oblivious to his owner's approval ratings, he strutted through the White House, somehow aware that in America, presidential pets are the closest thing we have to royalty. In 2013 he was joined by a second dog, Sunny, who shared this understanding. Free from responsibility, Sunny bounded recklessly and put his paws all over everything, like Prince Harry before he settled down.

Now that's the life, I thought. Give me the juiciest parts of 1600 Pennsylvania: the access, the ego boost, the extra spring of confidence in my step. But spare me the rest of it. Why do I have to lose sleep over the future of universal pre-K? Why does the thought of ending clean-energy tax credits feel so personal it balls my fists in rage?

I didn't even have to become a dog to be happy. I could be one of the giant spiders that infested the press office, where they got to be, and to eat, the flies on the wall. Or the red-tailed hawk who nested above the South Lawn, the one a group of fourth graders named Lincoln. Lincoln didn't understand the concept of "administration in crisis." Lincoln didn't know how it felt to wonder if years of effort were in vain. To him, the White House was nothing more than a giant squirrel buffet.

Nor did Lincoln know that presidents are expected to hold press conferences after humiliating defeats. But POTUS did. The day after the midterms he summoned reporters for Act I of the traditional post-shellacking script. Their questions left no doubt as to the narrative: "Defeated President Abandons Goals."

> "Do you feel any responsibility to recalibrate your agenda?"
>
> "Why not pull a page from the Clinton playbook and admit you have to make a much more dramatic shift in course?"

This was the part where President Obama was supposed to publicly scale back his ambitions. But something strange happened. Instead of humbling himself before the press corps, POTUS stood his ground. "The principles that we're fighting for, the things that motivate me every single day and motivate my staff every day—those things aren't going to change," he said. For those who don't speak press conference, let me translate. The president told reporters to fuck off.

It wasn't just talk. A few days after the presser, Beijing and Washington signed a joint climate deal, the first time China had ever agreed to limit carbon emissions. Coal-state senators like McConnell howled, but were powerless to stop it. That same week, I wrote a script for a Facebook video endorsing net neutrality. Cable companies whined fiercely, as did many of the lawmakers who counted on their donations. President Obama didn't care.

Yet climate and the Internet were mere warm-ups. No executive action was more fraught than immigration.

After the 2012 election, when Mitt Romney received just 27 percent of the Latino vote, Republicans conducted an autopsy of the results. It seems almost unbelievable today, but their most urgent recommendation was unanimous. Support immigration reform! Do it now!

GOP senators from states like Florida and Arizona got the message. Together with Democrats, they passed a bill to allow eleven million undocumented immigrants to become citizens over time. But in the House, where the Tea Party ruled and the districts were gerrymandered, reform went nowhere. In July 2014, the president announced he would no longer wait for Congress. He would act without delay.

Then, for several months, he delayed. This time the problem wasn't Republicans. It was Democrats. Already afraid to be tied to the president, skittish candidates begged him to hold back until after they won. He held back. They lost anyway. At the press conference the day after the midterms, reporters asked President Obama no fewer than five immigration questions, all of which were essentially the same. "Do you really want to go through with this?"

He did. The White House scheduled an address to the nation for November 20. I asked if I could write the speech. Cody agreed to give me a shot.

It was, by far, the most high-profile set of remarks I had ever written—a prime-time address, carried live on national TV. For the next two weeks, I fluttered about the West Wing: the Domestic Policy Council, the White House Counsel's Office, the Office of the Chief of Staff. Mostly, however, I poured over reams of survey data compiled by David Simas and his team. Night after night, they surveyed American homes, just as Joel Benenson had during the campaign. But where Joel's polling asked about a candidate, Simas asked about an issue.

"So, America's immigration system is broken. How does that make you *feel*?"

In the weeks before the big address, our research focused on voters we called Up for Grabs, or UFGs. UFGs were not exactly a melting pot. Overwhelmingly white. Politically independent. Predominately female. Typically suburban and middle-aged. It is a generalization, but only a slight one, to apply the name *Karen* to the entire group.

Karen wasn't against immigration entirely. That said, she believed in playing by the rules. She bristled at the thought of people who were coming here "the wrong way." With this in mind, I cast the president's actions as a kind of tough love. We would surge security along the southern border. We would force five million undocumented immigrants to emerge from the shadows and learn English. We would make them pay back taxes and a fine. Only then would we let them stay. Tough, cold-blooded, and couched in the language of self-interest, my draft was engineered to win UFG support. I sent it to Cody, certain I had earned a *boom!*

But Cody knew something I didn't. A regular at the 7:30 A.M. meeting, he had noticed a change in his boss.

"Why don't we go up and talk to him before he sees this?" he suggested.

POTUS time was scheduled. We arrived in the Oval, taking seats opposite the president at his desk. He reached toward my printed draft. Before he could lay a hand on it, however, Cody jumped in.

"I figured you wanted to focus on values for this one. Make the big, moral case."

"Exactly," said POTUS. "Let's go big here."

For the next several minutes, President Obama described his vision in terms of sweeping principles, not cost-benefit concerns. A system that left millions of workers unable to earn citizenship was unfair. A system that separated families was unjust. A system designed to reject rather than welcome immigrants was un-American. POTUS reached the end of his outline. Then, almost as an afterthought, he pointed to my draft on the desk.

"So Litt, I assume everything I said is in here?"

"Oh, yeah," I lied. "Totally."

We left the Oval, Cody making certain to retrieve my remarks so that POTUS wouldn't learn the truth. I didn't need to ask. I was getting bigfooted. A few nights later President Obama stood in the East Room, live on television, to deliver a completely rewritten speech.

"For more than two hundred years, our tradition of welcoming immigrants from around the world has given us a tremendous advantage over other nations.

"Are we a nation that tolerates the hypocrisy of a system where workers who pick our fruit and make our beds never have the chance to get right with the law?

"Scripture tells us that we shall not oppress a stranger, for we know the heart of a stranger—we were strangers once, too."

It's not fair to say President Obama ignored UFGs entirely. But it is fair to say their short-term preferences commanded less of his attention than ever before. Unlike Karen, a nineteen-year-old born to undocumented parents in Tucson or Reno wasn't up for grabs. She may not even have voted in 2014. But she was American. Obama was her president, too. And in what he called "the fourth quarter," he focused on building a country that delivered on its promise for everyone, voter and nonvoter, UFG and non-UFG alike.

The centerpiece of the president's announcement—a plan to protect five million undocumented immigrants—would eventually be overturned by the Supreme Court. But the change in tone was permanent. The 2015 SOTU was less than two months after the immigration address, and as it drew closer, Cody went into full warrior poet mode, cloistering himself in his office for weeks. The day of the speech, I finally saw him. He looked as though he'd been subsisting on a diet of rainwater and Nicorette.

"Are you going?" I asked.

"Nah," he said, exhausted. "You want to?"

Of course I wanted to. That night I loaded into the staff van, and the motorcade raced down Independence Avenue toward the brightly lit Capitol dome.

I had moved to Washington seven years ago, but this was my first time setting foot on the floor of Congress. I felt a strange mix of awe and fury. There was Mitch McConnell! (Actually much handsomer in person.) There was John Boehner! (Actually exactly the same.) I squeezed against a wall, identifying faces of congressional Republicans like a bird watcher who deeply loathes birds. There were so many of them. And so few of us.

If President Obama felt cowed to see a record number of critics in the audience, however, he didn't show it. Defiantly, proudly, he ticked off his accomplishments. Job creation at the fastest pace since the nineties. The highest-ever number of college graduates. The end of the combat mission in Afghanistan. In previous years, we had often caveated our achievements, not wanting to offend those who felt progress wasn't coming fast enough. This time, there was no pussyfooting around. "The State of the Union is strong," POTUS declared, in the very first page of the speech. Believe it or not, it was the first time since becoming president that Barack Obama had read those words aloud.

But the president saved his most memorable moment for the speech's final minutes. "I have no more campaigns to run," he announced. The line was supposed to be classy and high-minded. To everyone's surprise, however, the Republicans broke into spontaneous applause.

For a moment, I thought President Obama would simply ignore them. Instead, he paused and nodded tersely. I had watched him long enough to know what was coming next.

Uh-oh. He's gonna say some shit.

POTUS looked directly at his opponents and grinned. "I know," he said, "because I won both of them." Republicans glowered. Democrats, so recently humiliated at the polls, went nuts. Our critics would say the president's comeback was beneath the dignity of his office. But as far as I was concerned, while no squirrels were eaten, POTUS was acting positively Lincoln-esque.

PRESIDENT OBAMA'S ATTITUDE WASN'T THE ONLY THING UNSHACKled. So was the way he got his message across. When I first received my blue badge, there were really only two communications tools available to the White House: interview or speech. True, we had a new-media office. Once a year POTUS chatted with YouTube stars. But we treated the Internet as an accessory rather than a necessity, lipstick rather than pants.

That changed with *Between Two Ferns*. From the moment the video aired, we wondered why we hadn't done something like it sooner. When February 2015 rolled around, and Americans had their second chance to buy insurance through Obamacare, there was no doubt we would promote the law online again. The only question was with whom.

The answer was a drug deal. Not literally. No narcotics were exchanged for cash. In the White House, rather, a "drug deal" referred to a communications bargain in which media appearances were swapped. In this case our dealer was BuzzFeed, a site that became famous for listicles that spread like bird flu online:

17 WAYS YOU KNOW YOU GREW UP ON A HOUSEBOAT

---

15 FERRETS WHO JUST CAN'T EVEN

That sort of thing.

In 2011, however, BuzzFeed developed highbrow ambitions. Poaching Ben Smith, an editor from Politico, the newly created BuzzFeed News produced some of the best coverage of the 2012 campaign. Now, a few years later, BuzzFeed News was a player, and looking for the ultimate badge of media cred. They wanted a sit-down interview with the president. We wanted a video plugging Healthcare.gov.

Hence the drug deal. POTUS and Ben Smith would meet in the residence for an on-the-record chat that met the highest journalistic standards. Then the president would cross the hall to the library. There he would join a separate division of the company, BuzzFeed Motion Pictures, to shamelessly promote his law.

Our press team had no problem setting up the Ben Smith interview. From the start, however, it was clear that BuzzFeed Motion Pictures and the White House did not see eye to eye. Kori Schulman, our deputy digital director, was assigned to safeguard the president's equities. Like me, Kori was in her twenties. But applied to us, *millennial* was an adjective. Applied to our viral-content counterparts, it was a full-time job. With the relentless enthusiasm of an Instagram star on vacation, they proposed an idea: force POTUS to try weird American foods, then film him when he got grossed out.

"We think it will break the Internet!" they announced.

Kori and I tried to explain that there are certain unwritten rules to being president, and that one of those rules is "Don't mock things voters eat." But our dealers were unfamiliar with the concept of unwanted attention, and coming up with a replacement theme for the video was a nightmare. Prisoner swaps have been negotiated in less time.

Even when we did settle on a suitably inoffensive concept—"Things Everyone Does that You Do Too"—our equities proved difficult to protect. No, Kori and I explained, POTUS would not vape on camera. No, the president would not pull underwear from the crack of his ass.

"Are you sure? We really think it will break the Internet."

Eventually, we reached an agreement: POTUS would do things that were silly without being embarrassing. Making faces in the mirror. Using a selfie stick. Shooting a pretend jump shot. Mispronouncing *February* as he delivered his health care plug. The day before the shoot, when BuzzFeed Motion Pictures arrived for a walk-through in the residence, I hoped that we were finally on the same page.

One look at our video's director disabused me of this notion. I'll call this person Manbun after his most distinguishing feature, the same way some people are called Charity or Faith. Also, I get that not everyone dresses in business attire. I really do. But wearing skinny jeans to a White House meeting? As a nation, can't we agree that this is treason?

Perhaps I could have forgiven Manbun's clothing if his personality hadn't so perfectly matched it. For several minutes, he preened about the library, resembling nothing so much as a human Bo. Finally, he turned to Kori.

"I'm thinking we film the basketball shot first, then the selfie stick, then the health care plug." He pursed his lips in a caricature of thoughtfulness. #Listening. "What do *you* think?"

Kori was quick to reply. "Actually, we should do the health care plug first. We need to remind POTUS why he's doing this. Then we can do the basketball shot, then the selfie stick."

Manbun struck another pose. #ReallyListening. He remained frozen for several seconds. Then he suddenly sprung back to life.

"Great! So we'll do the basketball shot first, then the selfie stick, then the health care plug."

This was not an accident. He employed this tactic repeatedly. (1) State an opinion. (2) Make a big show of asking for feedback. (3) Restate the exact same opinion word for word.

It speaks to Kori's character that at no time did she kick Manbun enthusiastically in the testicles. She didn't even forcibly shave his head. And that, of course, is the essence of being a staffer. You deal with nonsense gladly, smile cheerfully at people who demean you, all so that your boss can worry about more important concerns. On the day of the drug deal, POTUS had no knowledge of the thousand small headaches his team had endured. After his straight-news interview, he bounded into the library, Hope Hall shadowing behind with her camera.

"Okay, what are we doin'?"

We began with the health care plug, a victory we had secured after hours of intense back-and-forth. As Kori predicted, starting with the serious stuff gave POTUS license to enjoy everything that came after. The basketball shot; the selfie stick; getting frustrated and saying, "Thanks, Obama," sarcastically: everything ran smoothly, almost eerily so. When our director indicated he was finished, POTUS invited the BuzzFeed Motion Pictures team to join him for a photo. Hope Hall dropped to one knee to film POTUS as he joked around with the crew. My jaw finally unclenched.

And then, just when I thought it was over, Manbun stepped forward. Reaching into his pants pocket, he retrieved a bright orange business card. Then he placed it in Barack Obama's hand.

"Here you go, Mr. President. Just in case you ever need more filming done."

I was stunned. Under any circumstance, Manbun's sales pitch would have been presumptuous. But with Hope kneeling right there on the floor, recording the entire exchange? It was unspeakably rude. From the corner of the room where I stood with other White House staffers, a small, collective gasp could be heard. President Obama merely smiled. He began heading for the door. I thought he would simply leave.

But he didn't. Instead, POTUS paused and turned directly toward his staff. When he spoke it was in a stage whisper, each word overflowing with disdain.

"Great," he said, gesturing with the orange card. "Maybe I'll put this in my *Rolodex*."

It is impossible, on the page, to capture the beauty of that moment. In fact, that's part of the beauty. Plausible deniability was preserved. But to those watching, there was absolutely no doubt what the president had done. It was the rhetorical equivalent of a body slam.

Nor was President Obama finished. Walking out the door, he turned to Marv, his aide.

"Man," he said loudly, shaking his head, "I just *love it* when people give me their cards."

Not all good presidents are good people. LBJ created Medicare and signed the Civil Rights Act while conducting his private life in a manner that would make a gorilla blush. But Barack Obama was the kind of person who noticed when a member of his staff was being insulted, and refused to walk away. He used his power to defend the dignity of others. He didn't have to do that. But he did.

I think this basic decency was reflected in President Obama's policies. I know for certain it's what convinced so many people to join his team, and remain there, even when times were hard. I still wasn't sure if POTUS could recover from the 2014 elections. I was only slowly rebuilding my trust. But as he left that library, with Hope Hall trailing behind him, a part of me loved Barack Obama more fiercely than ever before.

(Also, I have it on good authority that later, off the record, he made fun of Manbun's hair.)

GIVE BUZZFEED MOTION PICTURES CREDIT: THEY DID INDEED BREAK the Internet. Their video received tens of millions of views. By the end of spring, thanks to Obamacare, sixteen million more Americans were insured.

That wasn't all. The unemployment rate, 9.3 percent when POTUS took office, was down to 5.5 percent. Deficits were shrinking. Dependence on foreign oil was falling. We moved forward on new climate rules. We began to normalize relations with Cuba, untangling one of the last vestigial conflicts of the Cold War. President Obama was through waiting around for Congress. He was getting things done.

Americans took notice. Our approval ratings, thirteen points underwater on Election Day, were by April back to an even split. This was not supposed to happen. According to the narrative, Obama's White House was in its final throes. Instead, we defied political gravity every day.

"What's the mood like over there?" people asked.

"Everyone's feeling great!" I said. And this time I was being completely honest. There is nothing more exhilarating than being part of something meaningful as it rises from the dead. It animates your present. No less important, it justifies your past. All the vacations skipped, weddings or birthday parties not attended, lucrative job offers turned down. Everything suddenly makes sense. I had never been so happy in the White House.

Nor, as March turned to April, had I ever been so excited for a White House Correspondents' Dinner. As the big day drew closer, a comedy-writer friend from college, Andrew Law, captured the moment perfectly with a joke.

After the midterm elections, my advisors asked me, "Mr. President, do you have a bucket list?" And I told them, well, I have something that rhymes with "bucket list."

Take action on immigration? Bucket. Let's go for it.

New climate regs? Bucket. Why not?

I wondered if POTUS would cut the joke. Swearing in public, even in pun form, was risky. Besides, I could still remember that Biden "big stick" line from 2012. But this was the fourth quarter, and President Obama made only one change. In the very last sentence, he crossed out the words *Why not?*

Bucket, read the new version. *It's the right thing to do.*

The rest of POTUS's edits were no less fourth-quartery. Next to a joke about the Koch brothers, he'd written in the margin: *Something sharper—tougher?* Another setup referenced a recent comment from Dick Cheney. The former VP had said Obama was the worst president of his lifetime. When I tried a singe-don't-burn punch line—"And here I thought we were friends!"—POTUS was unimpressed. *We should come up with something sharper and more cutting here,* he wrote. Then, for emphasis, he added:

It's Cheney!

The president's energy was contagious. Jokes seemed to flow naturally that year. But that didn't mean the process was easy. We struggled to find a set piece, an unmistakable highlight of the night. How could we meet the moment? What could we do that we hadn't done before? I was on the verge of giving up when I remembered a holiday party.

Staffers like me crashed White House parties with two main objectives: stealing food and stalking celebrities. One night in 2014, I accomplished both. After pursuing my quarry past the eggnog and gingerbread houses, I cornered him between the lamb chops and the cauliflower mac and cheese.

"I write jokes for the president," I said. "And I'm a huge fan of *Key and Peele.*"

I was lucky. Keegan-Michael Key was not just a star of Comedy Central's most popular sketch show. He was also the most extroverted person I had ever met.

"Wow, thanks for saying hi!" he said, in a tone that suggested he actually meant it.

"POTUS loves your show, too. Maybe we can find a way to work together?" I got Keegan's e-mail and sent a follow-up. Then I promptly forgot the whole thing.

I wasn't lying, however. POTUS really was a fan. In particular, he enjoyed the recurring character of "Luther, Obama's Anger Translator." On Comedy Central, while his costar Jordan

Peele impersonated the president with trademark calm, Keegan jumped in after every sentence to rant about the nonsense America put him through. Each year I had taken the lead on the Correspondents' Dinner, we quietly wondered if we could put POTUS and Luther together in real life. But the timing was never right. In 2012 we weren't eager to suggest that Barack Obama was secretly an angry black man. In 2013 we were only two weeks removed from the Boston bombings. In 2014 we needed to stay humble after Healthcare.gov.

In 2015, though? Bucket. I fished Keegan's e-mail from my inbox.

A few days later, I heard back. Luther was a go. I threw together a script. Lovett punched it up from Hollywood. The next day, Cody and I went to the Oval to show the president our draft. There was no need for POTUS to practice his anger translator's lines, but he read them anyway, relishing the chance to vent.

"Y'all are ridiculous!" he told an imaginary press corps, swiping his finger through the air. He had penciled in that part himself. He added another joke as well, a few paragraphs later, about the media's coverage of the ebola epidemic.

What was that—one of the fifteen times you declared my presidency over?

Back in school, when my improv comedy troupe warmed up before a show, a sense of bulletproof authority would sometimes fill the air. We could see the future, and the future was awesome. What was true in a college common room was no less true in the Oval. As President Obama reached the end of his script, there was the quiet, bubbly feeling that comes from being on the verge of something special.

"Do you want us back here tomorrow morning?" asked Cody.

Ordinarily we did one last run-through around noon on the day of the speech. This time, however, POTUS shook his head.

"Nah," he said. "The truth is, I'm pretty fucking good at this."

THE FOLLOWING AFTERNOON, I SMUGGLED KEEGAN INTO THE WEST Wing, hiding him in Cody's office to avoid journalists' prying eyes. Finally, when we were sure the reporters had left for pre-parties, the two of us went to the residence to rehearse.

A podium was set up in the Map Room, the same place POTUS had practiced his immigration speech five months before. But this time the president was in a tuxedo and a far more relaxed mood. He and Keegan chatted as though they'd been doing shtick for years.

There was, however, a problem. The commander in chief couldn't keep a straight face. "Hold on to your lily-white butts," cried Keegan. The line, one of Lovett's suggestions from Los Angeles, was Luther's first of the night. POTUS burst out laughing.

"Okay, okay. I've got to keep it together."

But this was a promise President Obama clearly couldn't keep. He lost it every time. "I'm only getting a little worked up," he warned. "For the real thing, I'm going for it."

It took us a half dozen stops and starts, but finally, we reached the last page of the script. Here, we had written in a twist. POTUS would begin discussing climate change deniers in Congress, and it would make him furious. He'd grow angrier and angrier, eventually getting so worked up that even Luther couldn't calm him down.

"That part won't be hard," he assured Keegan. "I really do get angry, you know." Then he thought for a moment.

"I've just got to keep it together."

But he still couldn't manage it. During the second run-through, POTUS was just as hopeless as before. And there was no time for more practice. POTUS and Keegan stood on either side of me while I jotted down their final edits. Then we jumped in the motorcade and sped toward the Hilton for the speech. President Obama took his seat at the head table. Keegan went to his hotel room to put on a gray suit and eight gold rings. I spent the evening in my usual fashion, bouncing around with nervous energy like a gas molecule. It was only by chance that I happened to be standing right behind the curtain when POTUS ducked backstage. He smiled and shook his head.

"I just can't break," he said.

I was surprised POTUS knew the comedy jargon for laughing in the middle of a scene. I was unsurprised, however, by what came next.

"So, are we funny?"

It was the question President Obama had been asking for years. For years, I'd stammered a reply. Now, though, I had the perfect answer. I thought about his pep talk in the Roosevelt Room five months earlier. I thought about his determination to write his own history, to speak his own language, no matter the obstacles in his path. I thought about the twenty months left in the fourth quarter, and grinning, I looked America's first black president in the eye.

"Hold on," I said, "to your lily-white butts."

14

THE BIG ROCK CANDY MOUNTAIN

Steve and I sat in the catwalk, as frosty as exes at a parent-teacher conference. Below us, Luther the Anger Translator strode onstage. Keegan hadn't been kidding when he promised POTUS he would go for it. His veins bulged. His eyes bugged. As he screamed his opening line, the one about butts, my eyes darted to the president.

Please don't break. Please-don't-break-please-don't-break-please-don't-break.

To my horror, I saw POTUS swallow a laugh. It looked like he was about to lose it. And then, an instant before the point of no return, something clicked. It was like a bicycle changing gears. The president's solemn expression snapped into place. When he continued, it was in the calm, backyard-trampolines tone he used for the weekly address:

"We count on the press to shed light on the most important issues of the day."

"And we can count on Fox News to terrify old white people with some nonsense!"

Luther was as loud as ever, but I no longer needed to worry. POTUS wasn't breaking. For the next five minutes he was flawless, his timing impeccable, his body language perfect. When the subject turned to climate change denial, President Obama's anger was, as promised, real.

"What about our kids?! What kind of stupid, shortsighted, irresponsible bull—"

"Sir!" Luther cried. "Whoa whoa whoa whoa whoa. Whoa. Hey!"

"What?!" POTUS snapped, and the Hilton ballroom went wild.

Outside the ballroom, the reaction was no less enthusiastic. At the 9 A.M. meeting the next Monday, Jason Goldman, the White House digital director, delivered the news: Our Facebook clip of POTUS and Keegan's performance had been viewed thirty-five million times. In just forty-eight hours, "Luther, the Anger Translator" had become the most popular government-produced video in Internet history.

I bring this up for two reasons, and the first is bragging. But the second is more important. On some level, every White House staffer is an alchemist. You arrive at the building full of faith in miracles, striving to craft something flawless and shiny from the leaden scraps of real-world events. Before long, however, you realize it's never going to happen. Anything involving the real world, no matter how well executed, is bound to be impure.

Then one day, if you're lucky, you're going about your business when a shiny, golden nugget appears as if by magic in your lap. It's one of the greatest gifts of public service: you get to be part of small miracles, long after you've stopped believing that miracles of any size occur.

Legacy items. That's the term we used to describe these golden moments. Sometimes we even knew what it meant. Taking out bin Laden was a legacy item. So was rescuing the auto industry, bringing troops home from Iraq and Afghanistan, or repealing "Don't Ask, Don't Tell." But just as often, we imagined our legacy with the starry eyes of a hobo describing the Big Rock Candy Mountain. We dreamed of a distant utopia, a sunny political paradise, where the credit flows like a waterfall and approvals stay sky-high.

We weren't there yet. With twenty months to go until POTUS left office, our place in history was far from certain. But inside the building, something had undoubtedly changed. President Obama's jaunty, let's-go-for-it attitude was infectious. We no longer felt like turtles in our shells.

Our growing confidence was matched by growing competence as well. That's not to disparage the early days: as White Houses go, Obama's functioned fairly smoothly from the start. Still, the longer POTUS ran the institution, the more we learned from our mistakes. After the Healthcare.gov disaster, we began "red-teaming" a growing number of big decisions, assigning designated cynics to guard against undiluted hope. Confronted with its lack of diversity, Obamaworld gradually became a place where rooms full of white guys were the exception and not the rule. Baby steps, I know. But these baby steps made us a unicorn among bureaucracies—we improved over time.

Somewhat to my astonishment, so did I. At the risk of sounding boastful, I had now gone two full years without angering a sovereign nation. Even better, the White House finally felt like home. There was no one moment when the transformation happened. I didn't burst forth from a cocoon. It was more like learning a language. You study, you practice, you embarrass yourself. And then one day someone cuts you off in traffic and you call them a motherfucker in perfect Portuguese.

Whoa, you think. *I guess I'm learning.*

It must be said that my newfound fluency in White House was less a matter of national politics than office politics. I now had enough friends in the First Lady's office to sneak into East Room concerts without being unceremoniously bounced. Thanks to a growing list of policy contacts, I was winning more battles with fact-checkers. I had even mastered the dark art of adding a single, wildly indefensible claim to a draft when I knew lawyers would be reviewing. That way they could feel virtuous about cutting something while leaving the rest of my speech intact.

I even knew about a kind of top secret, commander in chief sonar: President Obama's incessant whistling. I'm not sure exactly when POTUS picked up this habit. Maybe 2014, maybe before. All I know for certain is that once he started, he couldn't be stopped. I'd be waiting for a photo line to finish or a taping to begin. Then, in the distance, I'd hear it, each note clear but the order random, like a child playing recorder or a bird badly botching a call. The louder the sound, the closer the president. When the whistling neared its crescendo, you knew to stand extra straight.

Like all things White House, the whistling also served an informal barometer of power. The more it annoyed you, the more time you spent with the boss. "It is *really* fucking irritating," announced one Assistant to the President in a hold room, cementing his place in the inner circle. His tone suggested he was dying to say something. But what?

"Excuse me, sir. I know your judicial nominees are stalled and Yemen's a nightmare, but could you please knock it the hell off?"

As a SAP, I wasn't important enough to be annoyed by the president's tics. I was, however, important enough to take outsiders to lunch at the Navy Mess. After showing us to our seats in the wood-paneled dining room, the uniformed wait staff would inquire about the signature dessert.

"Would anyone like a chocolate freedom?" At this, even people who didn't like chocolate perked up. "Chocolate freedom?" they asked. Before the server could answer, I jumped in.

"Excellent idea! Chocolate freedom for everybody!"

Later, molten fudge oozing from their lava cakes, vanilla frozen yogurt secure in its candy shell, my guests' faces would glow with rapture. There was no better ego trip. They were having a once-in-a-lifetime experience; I was having lunch. Drunk with power, or at least familiarity, I traipsed through the residence and self-righteously flashed my blue badge and was as cocksure as Rob Lowe in *The West Wing*. And then, just when I reached peak swagger, a coworker found me in my underwear in the coat closet of Air Force One.

Allow me to explain.

THERE ARE TWO KINDS OF MISFORTUNE THAT CAN BEFALL A WHITE House staffer. The first is the act of God. The van hits a pothole on the way to Andrews Air Force Base; ergo, you have coffee on your shirt.

The second is the mountaineering accident. There's no one big mistake. Rather, there's a series of small ones. An extra layer left at base camp. A carabineer improperly tightened. Boots a half size too small. No single oversight is worth mentioning. But add them up, throw in some bad luck, and before you know it you're standing before your colleagues on the presidential aircraft without pants.

The events that led to my mountaineering accident began the first week of June. POTUS was scheduled to fly from D.C. to Germany on Saturday, participate in a G7 summit on Sunday, and return the following afternoon. Cody and Terry had no interest in a thirty-six-hour jet lag fiesta. They assigned me the trip.

As it happened, this was exactly what I had hoped for. I was, by now, no stranger to Air Force One. I had taken more plane selfies than I could keep track of. I knew the best beer available was Yuengling Black & Tan. I had seen the sequels to both *Anchorman* and *Hot Tub Time Machine* on the in-flight entertainment system, and monitored at least one eBay auction from midair. But for all my time on board, my farthest trip had been California. I had never traveled overnight.

Which led to my first mistake: I didn't think about sleepwear. By the time Luke Rosa, our trip director, reminded us to bring something to change into, it was too late. I had no time to shop. The best I could do was rifle through my dresser, where two options emerged. The first was boxers and an oversize T-shirt. The second was a purchase from my freshman year of college, pajama pants adorned with pictures of the Incredible Hulk. It was a Sophie's choice of sleepwear. I chose Hulk.

My second mistake came just a few minutes after takeoff. We were in the staff cabin when, without warning or explanation, a member of the medical team entered and began handing out sleeping pills like candy. Foreign-trip veterans knew the drill. They immediately gobbled their meds and staked out prime nesting spots on the carpeted floor. But I stayed in my chair and abstained.

This is how I learned something: Air Force One is a surprisingly shitty place to sleep. It's cold. It's loud. The seats don't recline past forty-five degrees. I understand that "Air Force One ruined my eight hours" is the epitome of a first-world problem. Still, couldn't whoever installed the top secret communications array have made the track lighting slightly less harsh?

Apparently not. With no chemical help, sleep escaped me. When I finally drifted off, it was for an hour at most.

I awoke somewhere in French airspace, and here I made mistake number three: I ate. I wasn't even hungry. But I remembered the flight attendant's words of wisdom—an army marches on its stomach—and eager to be a good soldier, I gorged. Eggs. Croissants. Jam. Fruit. Coffee. Diligently plowing through my tray, I was too busy to notice the line forming along the plane's port side. By the time my plate was clean, a dozen people stood waiting to enter the lavatory and change into business attire. It would take at least an hour before my turn came. By that time, the plane would be on the ground.

This is the point when I panicked. Waking nightmares began cartwheeling through my head. *What if the plane lands, Angela Merkel is there to greet us, and I walk out onto the tarmac in my jammies?* In desperation, I scanned the staff cabin for a place to change. The small room with the computer and printer was out—it had no doors. Throwing a blanket over my chair was impossible; the risk of being noticed was too great. Then, suddenly, a stroke of genius! At the front of the cabin was a shallow closet, about six feet high and two feet deep, where staffers hung their coats.

It was the perfect plan. The moment my coworkers' heads were turned, I scooted into the closet and slid the door closed behind me. With lightning speed, I removed my shirt, socks, and pants. I reached for my trousers. I slid them off their hanger. And that's when Luke, the trip director, decided to retrieve his coat.

How will I be remembered by those who knew me? You can't help but ask yourself this question, especially if your work is part of something significant and grand. I liked to imagine that former colleagues, upon hearing my name many years from now, would picture someone self-assured and writerly. "He wrote those jokes I liked," they might say. Or, "Isn't he the one who made infrastructure finance come alive?"

Maybe some people really will remember me that way. Maybe not. But as I stood somewhere over Germany, looking from Luke's stunned expression to the giggling faces behind him, one thing was certain. For a small but significant portion of my colleagues, I will always be startled, pasty, and half naked, a pair of balled-up Hulk pajamas at my feet.

MY LEGACY, IN OTHER WORDS, WAS COMING INTO FOCUS. BUT HOW much did my record—my modest achievements, my less-than-modest humiliations—affect that of the president himself?

According to those responsible for boosting morale, the answer was "A lot." One day during the fourth quarter, a motivational poster appeared on the first floor of the EEOB, just outside the steps to West Exec. It was a black-and-white picture of Martin Luther King Jr. and Lyndon Johnson. The caption underneath was printed in large, blocky letters.

MEETINGS OF CHANGE

I always thought this was cheating. A more accurate "meetings" poster would have shown a staffer browsing Amazon for Christmas presents while listening to a conference call on mute. Still, I got the point. Everything I did somehow affected the Oval. Every action bolstered, or diminished, the most powerful person on earth.

Too bad the message on the walls was belied by the one in the floor. Throughout the EEOB hallways, the black-and-white tiles were dotted liberally with fossils. These were not impressive specimens, dinosaur teeth or mastodon tusks. What we had were mollusks out the wazoo. Most were little more than leggy spirals. A few looked like roaches. Tyrannosaurs these were not.

And yet I couldn't help but give our long-extinct friends a backstory. As far as I was concerned, the petrified little creatures beneath me had once been the White House staffers of their era. Swimming self-assuredly through prehistoric ooze, they reveled in their outsize importance. Perhaps they even thought about their legacies. Now, a few million years later, here they were, stomped on by the apex predators of a different age.

More than anything else, this cognitive dissonance—the disconnect between floor and poster—was the defining feature of White House life. There is something fundamentally ridiculous about putting humanity's fate in the hands of mortals. I'm not saying there's a better way to do it. I'm just saying that being a Very Important Person on one hand and a future fossil on the other takes an emotional toll. To live in such a contradiction, even for a righteous cause, is a low-grade form of agony. It pulls you apart at the seams.

Every White House staffer dealt with this inner turmoil differently. Some became short-tempered. Others took up yoga, smoking, or both. An impressive handful ran marathons. Almost everyone drank.

A few—fewer than you might think—became grandiose. These were the people whose egos ceased to exist except in relation to the Oval. They lost the ability to distinguish between themselves and the president, between petty personal jealousies and weighty national concerns. I don't blame those who came to believe their jobs made them more than human. A demigod complex is the malaria of the D.C. swamp. Still, it was sad to see good people fall victim.

And what about me? Did I ever succumb? By definition, I can't honestly say. Awash in paradox, it's possible (likely, on some occasions) that I acted like an asshole in ways I still don't realize. Here's what I know for sure. On days I didn't exercise, one beer was rarely sufficient to calm my nerves. More often than I care to admit, I went home and acted bratty until Jacqui felt as stressed as I did. There were even moments, often involving airline customer service, in which I thought without irony, *Don't you know who I am?*

Also, like everyone, I aged. No matter how young or old, junior or senior, White House service was measured in dog years. Between ages twenty-four and twenty-eight, my gray hairs went from curiosity to invasive species. The dark circles under my eyes became, essentially, tattoos. It only added insult to injury to hear the wrinkles emerging along my mouth referred to as "smile lines." I knew plenty of smiley people who didn't lose sleep trying to describe the budget, or furrow their brows each time they saw a member of Congress on TV. These people had hobbies. They went to brunch. Their mouths were fine.

Of course no one aged more, or more publicly, than POTUS. In 2012, I wrote a joke about how he would look like Morgan Freeman by the end of his second term.

"That's not even funny," he said.

Three years later, I could see his point. It wasn't just the well-documented explosion of gray hair. By 2015, President Obama's puffy eye circles had their own puffy eye circles. His smile lines appeared to have been carved by glaciers. His fingers seemed more delicate, his skin somehow thinner. To adjust to his aging eyes, speechwriters quietly raised the font size on printed remarks from 24 to 26. Every crisis and decision had become part of him, like rings inside a tree.

So, was it worth it? This, I think, is the essential question facing anyone with a tough job, and President Obama's was the toughest of all. He had chosen his profession with eyes open. He hadn't taken nearly as much risk as, say, a fighter pilot or marine. Still, the sacrifices were real.

And at the beginning of his second-to-last summer in office, it was also entirely possible they would be in vain. The unique appeal of Obama—what separated him not just from other politicians, but from other Democrats—was the promise that government could do more than fix problems. Government could be a vehicle for our common aspirations. Government could tackle age-old challenges, leaving our country fundamentally better than before.

This is why so much of POTUS's legacy came down to two persistent stains on the nation's soul. The defining policy issue of the Obama years was health care. The defining moral issue was race. And in June 2015, on each of these issues, progress was very much in doubt.

With health care, doubt came in the form of a lawsuit. By now, large parts of the Affordable Care Act were succeeding. The uninsured rate had never been lower. The cost of medical care was growing more slowly than before. Obamacare had survived a Supreme Court challenge, a government shutdown, and approximately a gajillion votes for repeal.

But the Holy Warriors refused to surrender. Undaunted, they tried another legal challenge, this time on even shakier ground. It was a truly silly argument—their case hinged on the lawmaking equivalent of a typo—and we figured the Supreme Court would never agree to hear it. Then the Supreme Court agreed to hear it. We began to sweat.

The second issue, race, was far more complicated. For obvious reasons, I don't claim to be an authority on the subject. But I am the world's foremost expert in pitching the following joke:

"No matter what else happens, I think there's a good chance I'll go down in history as America's first black president."

Year after year I tried to add this line to a monologue, and while it wasn't my finest work, I don't think that's why it got cut. In Obamaworld, there was always the nagging fear that the president's racial legacy would be defined by his skin color, his election, and nothing else. It was the kind of thing you didn't joke about.

Yet halfway through 2015, that worst-case scenario also appeared to be the most likely one. Deaths of African Americans at the hands of police brought decades of simmering tension to a boil. Black Lives Matter, a movement led by a new generation of civil rights activists, was rightly dissatisfied with the pace of change. After each tragedy came a wave of protests. Occasionally there were riots. Always there was the hope that this nightmare would be the last. But it never was. The tragedies piled up.

Then came Charleston. On June 17, a white, mop-topped twenty-one-year-old walked into a black church and, after taking part in Bible study, shot nine people dead. The killer's name was Dylann Roof, and he didn't choose his target at random. He hoped to start a race war or, even better, convince America a race war had already begun.

It seemed entirely possible he would succeed, setting off a perpetual motion machine of chaos, violence, and hate. And even if Roof's act of terror failed to metastasize, there was no question Charleston left America exhausted and depressed. On the next evening's *Daily Show,* Jon Stewart told his audience he was out of jokes. Instead, he delivered a deeply pessimistic tirade.

"We still won't do jack shit," he said. "Yeah. That's us."

The following Monday, a sense of gloom and uncertainty hung over the country. The White House was no exception. It was quite possible that, sometime before the week was out, the Supreme Court would rob millions of their health care. It was certain that on Friday the president would go to Charleston, where even the consoler in chief seemed no match for the cold logic of despair. Our string of fourth-quarter wins suddenly felt less exhilarating. Would President Obama's faith in America prove unfounded? Would his legacy, and ours, fall apart? No one knew.

And then, in a period of less than forty-eight hours, everything was answered.

IT STARTED ON THURSDAY, WITH A RULING BY THE SUPREME COURT. Lots of people think that White House aides get advance notice of these decisions. Surely someone knows a guy who knows a guy. No, they don't. Instead, judgments are unveiled like Oscar winners, a tradition both cruel and unusual for staffers who spend every waking hour creating the illusion of control. Imagine a careful, calculating poker player pausing halfway through a hand for an interlude of Russian roulette. That's what decision days are like.

In June 2015, these tense moments were even tenser than usual. At approximately 9:55 A.M., as the justices prepared to release their rulings, the entire building stopped whatever it was doing. The minutes ticked down, then the seconds. By now Jacqui was covered through her employer, so her insurance didn't hang in the balance. But for Zoe Lihn and millions like her, the next moment could be the most important of their lives. I bit my cheek nervously. My feet twitched.

And then, at 10 A.M., an anticlimax. The court would rule on something other than Obamacare. My adrenaline still pumping, I'd return to work.

Until June 25, that is. At 9:59 A.M. that Thursday, the Affordable Care Act was a Schrödinger's cat, simultaneously alive and dead. Then, just one minute later, the court unveiled its decision. By a 6–3 vote, the Holy Warriors were defeated. Barack Obama's signature achievement would not be overturned in court.

Long before the ruling, Cody had drafted multiple sets of remarks, one for each possible judgment. That morning, standing in the Rose Garden, POTUS delivered the most triumphant of the bunch.

"Five years in, this is no longer about a law. This is not about the Affordable Care Act as legislation, or Obamacare as a political football. This is health care in America."

Obviously, this was optimistic. Obamacare's days as a political football weren't exactly over. But the president's broader point was undeniable. For a population nearly twice the size of Virginia, the Affordable Care Act was not an ideological minefield or political prize. It was insurance. Congress could still weaken the law. Maybe they could even one day repeal it. But the fundamental principles behind Obamacare—that everyone deserves access to health care, and that the government can help secure it—were now woven into American life.

For the rest of the day, it was as though the entire White House had taken political ecstasy. Glowing coworkers smiled at each other for no reason. Meetings began and ended with hugs and high fives. Even the chocolate freedom tasted sweeter. It seemed nothing could top that moment.

Until, just one day later, something did.

Two somethings, actually. At 10 A.M. on Friday, June 26, I heard a wave of squealing outside my office. When I opened the door I found interns flooding the corridors, like the children in *Matilda* when Miss Trunchbull gets her due. The Supreme Court had just handed down another ruling. Same-sex marriage was legal nationwide.

I could barely believe it. The college students bouncing giddily in the hallways were not quite old enough to remember 2004, when opposing "the homosexual agenda" helped vault George W. Bush to a second term. But I was. A freshman in college, I was home for Thanksgiving that year when my childhood friend Chris came out to me. I will always remember what I thought next.

That's too bad. He'll never be able to get married.

This, remember, was in Manhattan. Even for a blue state like New York, legalizing marriage equality seemed not just unlikely but preposterous. You might as well have told me we'd all be riding dragons or growing prehensile tails. Now, just a decade later, Chris could get married in any state he pleased.

This time it was Sarada Peri, a fellow EEOB speechwriter,

who had written multiple versions of a Rose Garden address. She, too, got to shred all but the most inspiring speech.

"Progress on this journey often comes in small increments," said the president. "Two steps forward, one step back, propelled by the persistent effort of dedicated citizens. And then sometimes there are days like this, when that slow, steady effort is rewarded with justice that arrives like a thunderbolt."

In 239 years of presidential history, it was hard to find a moment like the one we were living through: a young century's most progressive law, and its most dramatic step toward equality, each ratified in a single twenty-four-hour span. Yet even now, President Obama had no time to celebrate the thunderbolts of justice. In just a few hours, he would be on his way to Charleston. He had a eulogy to give.

THE PREVIOUS WEEK, AS CODY BEGAN WORKING ON THE REMARKS for Charleston, I heard through the grapevine that POTUS wasn't excited about speaking. He had already addressed the nation eight times after mass shootings. Over and over, he said something would have to change. Then nothing changed. Why would this time be different?

By the day of the speech, however, the president had his answer. When Dylann Roof appeared in court, victims' family members, who had every reason to hate him, offered words of forgiveness instead. Rather than defend the Confederate flag flying over the South Carolina capitol, the state's Republican governor, Nikki Haley, called for it to be taken down. A killer had hoped to summon the worst of America through his actions. Instead, the best of America rose up in response. In his eulogy, President Obama found a word to explain what had happened.

Grace.

"As a nation, out of this terrible tragedy, God has visited grace upon us," the president said. "For he has allowed us to see where we've been blind." He spoke eloquently of these new insights. The pain caused by the Confederate flag. The horror of gun violence. The pernicious, lasting scars of racism.

Even more important than his words, however, was the way he said them. New speechwriters, writing their first remarks for an African American audience, would often be told, "Take 'em to church." For these speeches, POTUS borrowed his cadence from the civil rights movement and generations of black preachers. Our job was to find language to match. But until Charleston, these were smaller, targeted events, not national addresses. When the whole world was watching, POTUS was a kind of crossover artist, as much professor as pastor, as much Kennedy as King.

Not anymore. "The church is and always has been the center of African American life," he said, "a place to call our own."

Not *their* own. *Our* own. So often, after a tragedy or injustice, it fell to President Obama to explain what black America was going through, to be a kind of anger translator in reverse. But this time, instead of describing the anguish of his fellow Americans, he joined them in their grief. With every word the president spoke, you could see his heartbreak building. Then, without warning, he paused, looked down, and shook his head.

Watching on the livestream, I was confused. I had studied President Obama for years. I thought I knew every gesture. But this I hadn't seen before. Was he about to cry? To walk away? For another moment, there was only silence. An arena of mourners held its breath.

Then, softly, the most powerful person on earth began to sing.

Amazing grace, how sweet the sound,

One of the pastors onstage laughed, delighted. "Sing it, Mr. President!"

That saved a wretch like me.

Behind him, a second pastor nodded and clapped. Before long, the entire arena was singing. Some cried uncontrollably. Some smiled broadly. Most did both. As POTUS came to the speech's crescendo, a church organist began improvising a riff behind him. When the president reached his final line, he made sure to emphasize the fourth-to-last word.

"May God continue to shed his grace on the *United* States of America."

With his right hand, he gave the podium a satisfied little thump.

THERE ARE RARE MOMENTS WHEN EVERYONE IN WASHINGTON flocks to the White House on some unspoken command. The night of the bin Laden raid was one such moment. Friday, June 26, was another. I met Jacqui after work, about a mile north of the building. Without having to discuss it, we joined the throng.

Then we reached Lafayette Park, and we stopped and stared in amazement. The White House was lit up in the colors of the rainbow. Up and down Pennsylvania Avenue, couples of every description were holding hands. They cheered. They laughed. They cried. Mostly they looked up, toward the house where their president lived, lost in something as close as you can get in politics to wonder.

Every legacy needs its defining moment, an image that lives forever in our minds. Barack Obama gave us plenty to choose from. Historians will debate. But on that summer evening, I made my choice. For a part of me, the president will always be singing, almost but not quite on key.

Amazing grace, how sweet the sound,
That saved a wretch like me.

In that golden, miraculous moment, he looks vulnerable and human. His hair is gray. His heart is breaking. But as he thanks a higher power for the chance at something more than wretchedness, his voice is crystal clear.

I once was lost, but now am found
Was blind, but now I see.

In less than two days, Barack Obama had secured his place in history. No, the problems he faced were not solved forever. The Affordable Care Act was still under attack. Race was still a fault line. Discrimination against LGBT Americans remained far too real. But I now lived in a country where health care was a right and not a privilege; where you could marry who you loved; where a black president could go to the heart of the old Confederacy and take all of us, every color and creed, to church.

President Obama had not just fixed an economy. He had not just ended a war. He had made America a better place than the one where I grew up. The country I lived in seven years ago, the country I lived in seven *days* ago, had been fundamentally transformed.

On January 3, 2008, a freshman senator told me that people who love this country can change it. For the next seven and a half years I hoped that promise was true, and worked to make it true, but never knew for sure. Now, standing outside the iron gates of the rainbow White House, I no longer had to wonder. There was still a long road ahead for people who loved this country. But could they change it?

Yes. We did.

15

THE FINISH LINE

And yet something wasn't right. The past two days had been a nonstop moment of triumph, an Obamabot's wildest dream. But I didn't feel triumphant at all. Earlier that morning, right after the ruling legalizing gay marriage, I ran into our speechwriting intern Chelsea. Chelsea had started at the White House only a few weeks earlier. Now she was grinning ear to ear, basking in the latest piece of history.

"Just so you know," I heard myself say, "it's not always like this."

Uh-oh, I thought. *I gotta get out of here.*

Burnout. That was the word everyone outside the building used, but it wasn't quite right. What really took place was a kind of emotional erosion, each intense, thankless workday another drip on the idealistic portion of the soul. I knew plenty of staffers who arrived at the White House awestruck and shiny eyed.

Eighteen months later, they sounded like convicts planning to break out of jail.

I was not yet a prisoner of 1600 Pennsylvania. I still loved my job. But I didn't always like my job, and the more time passed, the larger the unlikable things loomed. I was tired of fighting with fact-checkers over sentences like "These steps are making a difference." I was tired of arguing with staffers who wanted POTUS to ack them by name at the Hanukkah party. I was tired of explaining that not every set of remarks needs its own, novel-length appendix on the budget process. These frustrations hadn't grown more numerous. But they had grown more frustrating.

Also, two years before my thirtieth birthday, I felt over-the-hill. "Of course I love *The West Wing*," gushed Chelsea the intern. "I watched every episode on Netflix in eighth grade." This was bad enough. The final straw came a few weeks later, when a young assistant in the real-life West Wing e-mailed me about a draft.

"*I don't think POTUS can say, 'We're all in this together,'*" he informed me. "*That's a line from High School Musical.*"

The benefit of seniority was that I felt no need to respond to such nonsense. I was secure in my knowledge that the concept of teamwork predated 2006. Still, even as I gained authority, I could feel myself losing perspective. Without meaning to, I had added a stop to my West Wing tours, directing guests to a glowing red cube near the Mess.

"This is our new soda machine!" I would announce. "The best part is you can combine flavors. I'm a big fan of raspberry lime ginger ale, but everyone has their favorites."

"Looks nice," my guests would murmur, eager to change the subject. "And what's that over there?"

"Oh, that? That's the door to the Situation Room."

SO YES, I WAS FEELING A LITTLE ERODED. EVEN SO, THE END OF THE second term was less than a year and a half away. I could do eighteen months standing on my head. In the end, it was the successes, not the frustrations, that made me feel ready to leave. POTUS himself put it best:

"We haven't won every battle. We've still got a lot more work to do. But when the cynics told us we couldn't change our country for the better, they were wrong."

To hear President Obama so firmly declare victory was satisfying beyond belief. But it also marked, as firmly as a graduation day, the end of one chapter and the beginning of the next. POTUS still had plenty on his to-do list. So did Cody: a final SOTU, a Democratic Convention speech, a farewell address. But with five first-rate speechwriters working beside me in the EEOB, and plenty more across Washington waiting in the wings, my own role no longer felt so vital. Thanks to the commissioned-officer seal on my business card, I was now essential personnel. But in the best possible way, I felt more nonessential than ever before.

And then a strange thing happened. Once I decided it was time to leave, I got way better at my job. I stopped worrying if drafts would get blown up. If I feared a line might go too far, I wrote it anyway. What were they going to do, fire me?

The week after the Charleston eulogy, Cody assigned me a speech on criminal justice reform, to be delivered at the national convention of the NAACP. I was excited about the topic, a cause whose time had come. But I could also hear echoes of the immigration address I had so badly bungled. For thousands of NAACP convention-goers, mass incarceration was a moral failing. It left gaping holes in families and neighborhoods, belying America's ideals. UFGs saw it differently. As far as Karen was concerned, the best argument was economic. Why spend eighty billion dollars a year on prisons when we can spend that money on roads and schools instead?

I sat in my office, knowing I would have to thread a needle and doubting that I could. But to my surprise, something had changed. I felt looser. Instead of writing in risk-averse fits and starts, I let the words and sentences flow: facts, stories, arguments, principles, wrapped around each other like strands strengthening a cord. When I sent the draft to Cody, I was certain it was the best of my career.

And then nothing. Hours passed. I waited for edits. No word. I had been writing speeches long enough to know what was happening. Cody was making major changes. When my draft came back, it would be soaked in red. Finally, long after I had abandoned hope, the light on my BlackBerry blinked. Eyes downcast, I opened my boss's reply.

Great job, bro! You really found the muse on this one.

I wasn't sure if Cody had only recently begun believing in muses, or if he had always believed in them and I had only now discovered mine. It didn't matter. I practically levitated.

The next day, Ferial Govashiri, the president's personal assistant, asked if I could swing by the Oval. For the first time in my four years working for Barack Obama, it was just me and POTUS in the room. We went through his notes, along with the paragraphs he had written by hand on a yellow legal pad. I returned to my office and made his edits. The next day we flew to Philadelphia, where, waiting backstage for the president, the board of the NAACP broke into a spontaneous rendition of "Amazing Grace."

During my time at the White House, it must be said that I wrote some truly bad speeches. Some were overly technical. Some were boring. A few were not just flawed but embarrassing, like poems you write in high school and find years later in a drawer. But on this particular afternoon, four years after my first remarks for the president, POTUS stood before a packed convention hall and delivered a perfect speech. The audience gasped when he described the scope of the problem. They applauded as he spoke of actions he would take. He went through cost-benefit analysis and detailed policy proposals without ever losing sight of his big moral case.

"Any system that allows us to turn a blind eye to hopelessness and despair, that's not a justice system, it is an injustice system."

It was the kind of line I never would have written a year earlier. It was too aggressive, too sweeping, too at risk of being labeled a sound bite. But now, in my own fourth quarter, I didn't care. Bucket. Why not?

President Obama finished the speech with an RP story. Born in Philadelphia, Jeff Copeland was arrested six times before age thirty-eight. To pass the time, Jeff used to spend hours jogging in place in his cell. Fellow inmates dubbed him "The Running Man."

Then one day, for reasons even he couldn't explain, Jeff decided to turn his life around. He got sober. He graduated summa cum laude from community college, with a 3.95 GPA. He found a job.

"Just two years ago," President Obama told the crowd, "the Running Man ran his first marathon—because he's going somewhere now."

There was a giant round of applause. Then POTUS continued:

"We are not perfect, but we have the capacity to be more perfect. Mile after mile; step after step. And they pile up, one after the other, and pretty soon that finish line starts getting into sight, and we are not where we were. We're in a better place."

"We are not where we were." My life was about as different from Jeff Copeland's as it is possible to imagine. And yet what applied to him also applied, in its own way, to me. I thought about the January night I first found my candidate, or the days I spent at the Crisis Hut, playing Minesweeper like it was my job. I was so far from the place where I had started. Now, my finish line was in sight.

I DIDN'T GO IMMEDIATELY. I WAS IN THE ROOSEVELT ROOM AT 9 A.M. on November 5, 2015, exactly one year after our post-midterm pep talk, when the door swung open and POTUS appeared. Plenty of folks had counted us out, he reminded us. But look at how far we had come. A nuclear deal with Iran. Millions of new jobs. Higher wages. More clean energy. Countless new protections for consumer rights, and civil rights, and women's rights, and workers' rights. I clapped as enthusiastically as anyone, knowing it was the last time I'd be in this kind of room with the president of the United States.

Cody and I decided on a date for my departure: Friday, January 22, 2016. As it drew nearer, everything about the building began to seem more specific, more place-y. The EEOB basement was at its murderiest. The grease smell outside Ike's smelled greasier than ever before.

No less acutely, I began to realize all the things I would miss. The perfume of flowers in the Rose Garden. The floor-tile fossils. The impressive-yet-humble men's room. The weighty significance of the door to the senior staff gym. It was a cliché, but a true one, that I would miss my coworkers more than any job perk. There are simply not many places where 95 percent of the people—even the ones who drive you crazy—are really, really good at their jobs.

What I would miss most of all, however, was power. I know I'm not supposed to say that. It makes me sound like a Bond villain. But I don't care. For a person who hopes to make things better, wanting power is no different than a singer craving a microphone, or an actor yearning for a stage. For five years, whatever gifts I possessed were amplified a millionfold by the eighteen acres where I worked. If I had a stroke of insight, or a dalliance with a muse, my entire country was in some small way better for it. Who wouldn't want a taste of power like that?

AS MY FINAL WEEK OF WORK BEGAN, A PART OF ME HOPED FOR A last-minute summons to the Oval.

"Litt! You can't leave. The country will fall apart without you."

No such summons ever came, of course. Instead of postponing my departure to take on an alien invasion or zombie apocalypse, I spent my last week in the White House filling out a dazzling array of forms. It was a scavenger hunt in reverse. Had I dropped off my gym key at the athletic office? Had my computer and BlackBerry been handed over to operations? Had I returned the library's FDR biography, the one I renewed no fewer than nineteen times? These questions took up most of Monday and Tuesday.

Wednesday, January 20, belonged to my final POTUS speech. This was for the trip to Detroit, the one that began with a two-thousand-calorie lunch, was punctuated by a snowstorm, and ended with the motorcade snarled in traffic. On Thursday, still shaken after the previous night's slip-slide through the District of Columbia, I began to pack.

It is fair to say I was not known for office tidiness. It is fairer to say that my work space was a dreadful, undeniable mess. Cleaning out my desk was akin to an archeological dig. Also, there was less time for tidying up than I had anticipated. After the relatively light dusting on Wednesday evening, one of the biggest blizzards in D.C. history was forecast to arrive on Friday afternoon. I crammed years' worth of detritus into boxes as quickly as I could.

Without meaning to, I was creating an extremely mundane time capsule. Several pounds of gym clothes, along with one open tin of shoe polish and seven mismatched socks. My Nerf gun, a secret-Santa gift from Cody, which I shot into the air like a warlord during fits of writer's block. Add to this the impressive collection of plastic utensils I'd pilfered from the cafeteria, and three dozen thick history books I meant to read but never did. I dragged the heavy cardboard boxes to the car, more grateful than ever for my parking pass. Even so, I was at it well into the night.

On Friday, my last day at the White House, the federal government was closed in anticipation of the coming blizzard. Employees were instructed to telecommute. But cleaning an office can't be done remotely. My only option was to return. This time I left the car at home but brought a suitcase, a wheeled one, big enough to carry several years' worth of junk. Office supplies. Mini bottles of rum and whisky. A single, lonely shoe. Mindful of the storm fast approaching, I stuffed everything in without regard to sentiment or preservation. Half a two-pack of ibuprofen. Boxes upon boxes of presidential M&M's.

It was only when I reached a worn, tarnished picture frame that I paused.

Inside was a note addressed to my great-grandfather. In 1934, he wrote a letter to the White House, wishing FDR a happy birthday. He never received a presidential reply. But he did get a thank-you from Louis Howe, one of Roosevelt's top aides, along with an assurance that his message had reached the president's desk. I never met my great-grandfather. Other than his name, Maurice, I know almost nothing about him. But I know how much that note on White House stationery meant to him: he framed it and kept it until he died.

What would he make of all this? I wondered. Imagine if he could see me, just three generations later, sitting a few hundred yards from the Oval, my own stack of White House stationery in my desk. In human history, what a vanishingly rare story. In America, how typical.

My next thought was less poetic. *Holy shit. It's fucking snowing.*

As forecast, a full-blown blizzard had arrived. The flakes were coming down thick. An uncomfortable thought struck me: if the snow accumulated more than a few inches, the wheels on my suitcase would be useless. I would be stranded on the sidewalk. I had to leave. Now.

Thus began my final, not-so-dignified stroll through the corridors of power. I grabbed the last few items from my office and laced my snow boots in a panic. Already sweating in my down jacket, I sprinted through the wide, empty hall.

The moment I stepped outside, however, I stopped to look around. I couldn't help it. In the falling snow, the White House campus and the outside world had reversed roles. Out there life was messy and chaotic. In here, everything was untouched, pristine, calm. I walked until I reached the north end of West Executive Avenue. I passed a security checkpoint and heard the familiar click of the lock sliding open in the gate.

One year later, when many of my colleagues left the White House for the last time, they wouldn't have the luxury of feeling wistful. So much of what defined Obama—so much of what defined America—would be under assault like never before. But I was lucky. Standing along the fence between the building and Pennsylvania Avenue, I could think back to what POTUS had said about Jeff Copeland near the conclusion of his speech.

"We are not perfect, but we have the capacity to be more perfect. Mile after mile; step after step."

As a small avalanche fell from my hat onto my jacket, I thought of all the miles that had piled up over the past eight years. I had knocked on doors and driven naked. I had organized a county and scrubbed Janice Maier's table till it gleamed.

I sang the *Golden Girls* theme song in the Oval. I watched a tiny man surf Jesse Jackson's coat. In a convention hall in Charlotte, I met a mom from Arizona who would never stop fighting for her little girl. I was disillusioned more times than I thought possible. I was reinspired more times than I could count. I navigated Healthcare.gov for a woman, the highest test of love. I helped break the Internet. I wrote one perfect speech. I found a salmon in the toilet and was caught half naked on Air Force One and told the president he looked like Hitler to his face.

I was, I felt, more perfect than when I started. As one chapter ends and another begins, what more can you ask for than that?

With my right hand I clutched my suitcase. With my left I gripped the iron bar. Blinking snow from my eyelashes, I took one last look at the building before pushing into the storm. And then, smiling despite everything, I walked out the gate, to America.

Epilogue

SQUISHING THE SCORPION

"So," she asks, "how's that whole hopey, changey thing workin' out for ya?"

It's January 28, 2017, and imaginary Sarah Palin is whispering in my ear. She's been doing this nonstop since Donald Trump became president eight days ago. I'll be going for a run, cooking tilapia with Jacqui, picking up a six-pack of paper towels from CVS. No matter what I'm doing, her folksy insult lingers, a kind of malevolent background hum.

Right now what I'm doing—what I've flown halfway across the country to do—is watch a six-year-old practice karate. She's tiny, even for a first grader. Along with her white uniform, she wears bright purple glasses. A bright purple hair tie holds her ponytail in place. We're in the kind of pint-size martial arts studio that caters to children: a few punching bags, some motivational posters, a long table piled high with Capri Sun. At the center of the mat stands the instructor, a young blond woman who goes by Miss B.

"Pretend I'm coming toward you," Miss B says. She leans down, protecting herself with a thick black practice pad. "Ready?"

The student's size masks surprising fury. "HICE!" she yells. (I guess this is how kids these days say "Hai-ya!") A golf-ball-size knee launches upward, hitting Miss B's practice pad with a satisfying *thwack*.

"HICE!" *Thwack*.

"HICE!" *Thwack*.

From our bench along one wall of the studio, the little girl's dad leans toward me. "She tried that on me once," he sighs. "I was standing in the bathroom and she ran up and got my leg." I laugh and try to imagine the scene.

But Donald Trump is president, and my imagination is permanently elsewhere these days. From the small Arizona karate studio, my mind wanders to a beautiful D.C. day in the fall of 2015. I'm returning from the West Wing to my office. As I walk up the EEOB steps, David Simas walks down. Beneath his cliffs of hair sits a bewildered, can-you-believe-it smile.

"Donald Trump!" grins the president's political director.

"Donald Trump!" I grin back.

Don't get me wrong—we're not happy to see him run. His campaign, only a few months old, is already a national disgrace. What Bush hinted at, and Palin masked with a smile, Trump bellows at the top of his lungs. Dog whistles have become primal screams: Mexicans are murderers and rapists. Illegals are stealing our jobs. Obama founded ISIS. Vladimir Putin is a role model. Journalists are the enemy. White supremacists are just fine.

Yet as campaign people, it's impossible not to revel in the chaos. Watching Trump tear through the GOP primaries is like watching a sworn enemy suddenly realize the full implications of owning a pet chimpanzee. Republicans had plenty of chances to cut him loose. They chose not to. Now it seems obvious that Trump is poised to tear apart the conservative movement. A majority of voters will reject him. There's no way he will win.

We are two-thirds right, which is a little like describing the *Hindenburg*'s final voyage as "mostly without incident." Still, it's worth noting: the party of Ronald Reagan died with Donald Trump. For years, Republicans like Paul Ryan pretended their voters cared about conservative ideology: tax cuts for rich people; widespread deregulation; manly chins. Trump exposed all that as nonsense. More than anything, he realized, the Republican base was motivated by a kind of equal-opportunity resentment. To the question, "Who's screwing you over?" Trump's answer was simple. Everyone. He attacked undocumented immigrants one minute, Wall Street bankers the next. He was buoyed by fears of global elites and racial minorities alike.

And he guessed, correctly, that many Americans felt democracy was a luxury we could no longer afford. "I alone can fix it!" he announced in his convention speech. In the hall, the audience cheered. On our gray Martha Stewart Living couch, Jacqui and I stared at our TV in shock.

"There's no way voters will actually go for that, right?" I asked.

"No," Jacqui said. "No way." But her voice lacked confidence.

Technically, of course, Jacqui was right: Hillary Clinton won the most votes on Election Day. But those votes came from the wrong places. Trump surged in the swing states—in my former turf of Wayne County, Ohio, he beat Mitt Romney's margin by a whopping thirteen points. By the end of the night, thanks to the Electoral College, he was president-elect.

That was three months ago. Now it's Saturday, January 28. In Washington, our new commander in chief is calling Putin from the Oval. In Arizona, Miss B goes over footwork before next week's yellow-belt exam.

"Imagine there's a bug on the floor," she tells her eager student. "What's the grossest bug you can think of?"

"A scorpion!"

"All right, a scorpion. After you kick, I want you to put your foot down and squish the scorpion, okay?"

The bright purple hair tie bobs enthusiastically. The first grader resumes practice. And as she does, I remember the first time I saw her: in a campaign video, just a few days old, tubes running in and out of her chest. The second time I saw her she was in her father's arms in Charlotte, watching her mother address a rapt convention hall. One year after that, thanks to the law that lifted the lifetime cap on her insurance, she had her third and final open-heart surgery. It went better than even Stacey had hoped.

"HICE!" *Thwack.*

"HICE!" *Thwack.*

Zoe Lihn brings her tiny foot down, hard, onto the mat. That scorpion never stood a chance.

THIS IS WHY I'VE COME TO ARIZONA. IN WASHINGTON, THEY'RE already talking about undoing everything Obama fought for. Which executive orders will be rescinded? Which laws will be repealed? And it's true: part of Barack Obama's legacy lives on paper. But another part goes to elementary school in Phoenix, loves Mexican food and stuffed animals, and is hoping to ace her yellow-belt test next week.

Some accomplishments, in other words, are impossible to undo. Millions of people found jobs thanks to Obama's decisions. Soldiers who spent Christmas 2008 in Iraq and Afghanistan spent Christmas 2016 at home. Bin Laden was on the loose eight years ago. Now he isn't. Even Trump can't change that.

Also, while POTUS didn't upend our politics like I had hoped he would, he completely transformed our culture. Barack

Obama grew up in a kind of gray area, torn between races and worlds. During his presidency, a new generation of outsiders—gays and lesbians, African Americans, immigrants, science nerds, kids with funny names, and so many others—grew up feeling part of America in a way they hadn't before. And a new generation of insiders—kids like me but younger—grew up learning there's nothing shameful or scary about people who are different. That kind of sea change can be postponed for a while. But it's nearly impossible to reverse.

Nor is repealing laws as simple as it sounds. I don't mean to be naive. As Zoe practices a series of decisive blocking motions, the Trump presidency is just eight days old, and already it's clear that plenty of progress will be rolled back.

But it turns out even voters who didn't like Obama shared many of his priorities. Americans want clean air and water. They care more about middle-class wages than upper-income tax cuts. People insured thanks to the Affordable Care Act have no interest in seeing their health care disappear. In a representative democracy like ours, popular opinion isn't everything. It is, however, something. And right now, it's largely on the former POTUS's side.

So, at the end of the day, was Barack Obama a good president? It's a question I thought had been settled the night of the Charleston speech. Now here I am, in the shadow of Trump, asking it again. And yet, while the future has become frighteningly uncertain, it's impossible to watch Zoe Lihn lose herself in a flurry of kicks and punches without being confident about the past. Of course he was a good president. Look at her go.

BUT ENOUGH ABOUT OBAMA'S LEGACY. WHAT ABOUT THE REST OF US?

I mean that sincerely. For some people in Obamaworld, POTUS included, the White House was the climax of a life story. But for many more it wasn't. Like thousands of my fellow staffers, working for the president was my first real job out of college. Our names don't belong in the history books. No professors will devote their lives to examining choices we made. Still, there are questions that need answers. Was going into government the right way to spend our twenties? What should the next generation do differently? Where do we go from here?

Whenever I ask myself these questions, I remember my college commencement speaker, former British prime minister Tony Blair. I don't recall a word he said, but I do recall thinking his speech should have been titled, "Amusing Things That Happened to Me, Former British Prime Minister Tony Blair." He riffled through anecdotes, made fun of the French, and told us to follow our hearts. Not exactly profound.

Yet without meaning to, I took his advice. I didn't enter government out of high-minded principle. I did it because, for a brief period coinciding with my graduation, there was no more exciting place to be. My heart blazed a trail to the bureaucracy. I followed obediently behind.

Which is how, entirely by accident, I chose wisely. My two terms in Obamaworld didn't convince me everyone should go into politics. But they did convince me everyone should go into public service. This is not a matter of job description. Instead, it's a matter of moral orientation, about regularly and honestly asking, "Am I doing enough good?" If you devote yourself only to yourself—if your heart strives above all else for fame, or money, or power, or even happiness—I'm not saying you're a bad person. But you are making a bad choice.

Choose service instead. Not because The World Needs You. It probably doesn't. The world will be fine, and if it won't be fine, you alone can't fix it. Choose service because there is nothing more insufferable than a talented, driven person who is also completely self-obsessed. Those people are awful. They spend their lives trying to fill a hole and digging it deeper instead.

Anyone can be successful. Only service can make you realize how insignificant—and yet how meaningful—your time on earth really is.

No less important, if you're a young person, public service will teach you stuff you didn't learn in school.

Eight years in Obamaworld taught me perseverance. As a twenty-one-year-old, I assumed people on the right side of history were accompanied by a serene, ever-present glow. I figured doing good meant feeling good. Otherwise, why bother?

Today when I think about the right side of history, I remember a colleague's going-away party. At the end of the evening, one of his mentors gave a toast.

"For an entire year, Jacob took data from memos and turned it into PowerPoints. Then he took those exact same PowerPoints, and turned them back into memos." After a pause for knowing laughter, the toast continued.

"And because he did that, 140 million American workers got a payroll tax cut."

I have no doubt both halves of this anecdote are true. Yes, change comes from marchers, and visionaries, and freedom fighters. But I know now that change also comes from office workers in stuffy buildings with ugly wall-to-wall carpet. Change comes from people who set a worthy goal, put themselves in position to achieve it, and keep working long after the warm and fuzzy feelings disappear. Barack Obama's election was a triumph of hope. But his presidency was a triumph of persistence.

Eight years in Obamaworld taught me focus. Each news cycle—already shrunk to twenty-four hours when POTUS took office—lasted mere seconds by the time he left. He faced constant pressure to approach every issue with the frantic, hair-on-fire urgency of a tweet. More than once, I found myself frustrated by the president's patience. To me it seemed more like delay. But nine times out of ten, Obama was right. The secret to solving big problems, I learned, is knowing which little problems to ignore.

The list of Things Obamaworld Taught Me could go on for several pages. I learned that decisions are only as good as the decision-making process. That generosity is a habit and not a trait. That all human beings, even presidents, look goofy chewing gum.

But here, beyond a shadow of a doubt, is the single most valuable lesson I learned in public service: There are no grown-ups, at least not in the way I imagined as a kid. Once you reach a certain age, the world has no more parents. But it contains a truly shocking number of children. These children come in all ages, in all sizes, from every walk of life and every corner of the political map.

And this is the reason I'm most grateful for my time in Obamaworld. For eight formative years, often against my will, I was forced to act like an adult. Children strive for pleasure; adults for fulfillment. Children demand adoration; adults earn respect. Children find worth in what they acquire; adults find worth in the responsibilities they bear. And more than anything else, what separates adults and children is the way in which they love.

I didn't always understand this. One night in 2011, during the first month Jacqui and I were dating, I had a few too many Island Juleps and began to ramble.

"The problem," I announced, "is that there's not enough love in our politics."

In hindsight, I had no clue what I was talking about. If you're a certain type of young person in Washington, this is simply what you say to someone you hope will sleep with you. I consider myself lucky it worked. But if pressed, I suppose I would have admitted that by *love* I meant something more like infatuation. It was the way I felt the very first time I saw that freshman senator from Illinois. *He's flawless! He gets me! Only he can make the world as perfect as he is!*

What I know now is that this kind of love, while wonderful, is for kids. Real love—for a president, for a person, for a country—is more textured than that. Real love is about fighting for something long after its flaws are laid bare. It's about caring so deeply, you have no choice but to place another's well-being above your own. Love is not a feeling. It transcends feelings. Love is what allows us to be disillusioned and to somehow still believe.

And love has a way of brightening even the darkest moments. Just four days after Trump was elected president, I got down on one knee and asked Jacqui to continue believing in me, no matter how disillusioned she became. I think I put it more romantically than that. It was kind of a blur. Regardless, I consider myself lucky she said yes.

IT'S A STRANGE THING TO SAY, BUT I SUSPECT OUR NEW CHIEF executive has no idea how I felt when Jacqui agreed to marry me. Along with his other flaws—or really, at the root of them—Donald Trump is a seventy-year-old kid.

By the time I visit Zoe Lihn in Phoenix, eight days after the inauguration, it's becoming clear that installing a toddler in the White House has consequences. Week one of the Trump presidency has featured countless rounds of a game I call "How much worse is America since I last checked my phone?"

The answer, usually, is "A lot." EPA scientists have been placed under a gag order. New deportation rules are designed to tear families apart. The rule of law is already fraying, and the Republican Congress is too busy targeting consumer protections to care.

It's terrifying. But it's not surprising. What *is* surprising is the number of people who love their country despite its flaws, and refuse to give it up without a fight. Friends who never before cared about politics are pouring out of the woodwork, asking what they can do. I don't have an easy fix for them. But their question, in and of itself, is an answer. It's the best of America: defiant, proud, optimistic in spite of everything. It reminds me of nothing so much as the 2008 campaign.

It goes without saying that Stacey Lihn is part of it. She tells me that Election Night was devastating. The morning after, she had trouble getting out of bed. But she never doubted she would return to the fight. She had no choice. On January 15, she stood outside her senator's office in Arizona and demanded that he protect her daughter's care. The day after the inauguration, she was one of three million people who joined Women's Marches across the country—the biggest single protest America had ever seen.

"It's kind of like our journey with Zoe," she tells me. "You get dealt a rough hand, and you have to rally and you have to come back."

Sitting at Stacey's kitchen table, I have so many questions. Does she think we'll win? What does she do when she gets discouraged? Has she, too, been unable to sleep? But our interview is cut short by a hyperactive six-year-old. Zoe has recently acquired a scooter. She's desperate to go outside and play. And that's the way it should be. Thanks to a man who lived in the White House—and to all the people who put him there—Zoe is as bouncy and impatient as any other kid.

That doesn't mean everything will be okay. Pulling out of the Lihns' driveway, I turn on my phone, play my new favorite game, and discover that America has become much worse in the hours since I last checked. President Trump is hoping to make good on his campaign promise to ban Muslims, starting with seven nations from Africa and the Middle East. The executive order he's just signed is a stunning blend of ineptitude and malice. Translators who served with U.S. troops in Iraq are being held like criminals at the airport. Green-card holders are being illegally detained. Babies can't get food. The elderly can't get medicine. The stories are heartbreaking.

But they're not finished. Within hours, thousands of fellow citizens spontaneously descend upon America's airports. Volunteer lawyers fan out to represent detainees. The ACLU takes the president to court and wins. Despite everything, I end the night smiling as I watch a video on my phone.

The footage is from JFK airport, the same airport I flew into the night Barack Obama's speech changed my life. This time, though, the terminal is packed. The energy is as intense as anything on the campaign. As the camera pans toward a security door, a middle-aged woman in a head scarf shuffles out of detention. Relatives run to embrace her. She looks almost dizzy with relief. As she slowly exits the airport, the crowd breaks into raucous cheers:

"U-S-A! U-S-A!"

To my surprise, quietly in my hotel room, I join in.

THE FOLLOWING AFTERNOON, STACEY LIHN GOES TO THE PHOENIX airport to protest, and I go there for my flight home. The plane takes off. I look out over the shrinking sea of cars and houses. Then, like clockwork, Sarah Palin chimes in.

"So . . ." she asks.

But for the first time in eight days, her question is pushed aside by a million more important ones. Will Zoe Lihn be able to keep her health care? Will she grow up in a land of freedom and opportunity? Will she achieve her dreams, which at the moment include earning a black belt and becoming a school principal so she can order fire drills whenever she feels like it?

I don't know. No one does. But I do know this. While Donald Trump may be our president, he does not define our country. I don't think he ever will. Zoe Lihn is six years old, and for now, this is still Barack Obama's America.

Anything is possible.

ACKS

All speechwriters know that it's important to keep acknowledgments short. They also know that, if you can't keep the acks short, you must under no circumstances accidentally leave anyone out. I'm about to break the first rule, and most likely the second as well. So with apologies in advance, thank you to the following people, without whom I could not have written this book.

My agent, Dan Greenberg, for understanding what I was trying to do before I did.

My editor, Denise Oswald, for asking me questions I had no answers to and sticking with me until I found some.

Ashley Garland, James Faccinto, Meghan Deans, Miriam Parker, Sonya Cheuse, Emma Janaskie, and the rest of Ecco, for making sure this book is worth reading and that you're reading it.

Catherine Burns, and the entire Moth family, for their friendship, encouragement, and storytelling genius.

Mike Farah, Brad Jenkins, and the team at Funny Or Die, for believing in, and making lots of, quality stuff.

Amanda Hymson and Jason Richman at UTA, for being strategic and supportive in equal measure.

Vinca LaFleur, Jeff Nussbaum, Paul Orzulak, and Jeff Shesol, for deciding I could write speeches when there was not much evidence to support that notion.

Valerie Jarrett, Cody Keenan, Jon Favreau, and Mike Strautmanis, for giving me the chances (and sometimes second chances) of a lifetime.

Everyone who wrote POTUS jokes—for no pay, and usually for no credit—while I was at the White House: Judd Apatow, David Axelrod, Beth Armogida, Kevin Bleyer, Jon Lovett, Andrew Law, Nina Pedrad, Pete Schultz, Nell Scovell, Rachel Sklar, Will Stephen, Katie Rich, Tommy Vietor, and the West Wing Writers crew. With so much extraordinary material, each monologue could have easily been five times as long.

My colleagues on the POTUS and FLOTUS speechwriting teams: Dave Cavell, Laura Dean, Sarah Hurwitz, Susannah Jacob, Steve Krupin, Tyler Lechtenberg, Kyle O'Connor, Sarada Peri, Aneesh Raman, Carlin Reichel, Megan Rooney, and Terry Szuplat. I'll always be grateful for your talent, your friendship, and for the fact that you only occasionally made fun of how messy my office was.

The people (and the following is a partial list) who gave up their time to make this book just a little bit better—to read a chapter, offer encouragement, or share an invaluable piece of advice: Mike Birbiglia, Joanna Coles, Billy Eichner, Ashley Fox, Peter Godwin, Ben Orlin, Tig Notaro, B. J. Novak, Eric Ortner, Kevin Roe, David Sedaris, Erik Smith, Kimball Stroud, and Alexandra Veitch.

And finally: To my Ohio volunteers, White House co-workers, and campaign colleagues, for inspiring me. To the friends I ignored while writing this book, for forgiving me now that I'm done with it. To my family, for being generous and kind. To my parents, for believing in me and always leading by example. And most of all, to Jacqui, for everything.